R. GUPTA'S®

CHEMISTRY

Formulae & Definitions

by

Ramanand Thakur
&
RPH Editorial Board

Ramesh Publishing House, New Delhi

Published by
O.P. Gupta *for* Ramesh Publishing House
Admin. Office
12-H, New Daryaganj Road, Opp. Officers' Mess,
New Delhi-110002 ✆ 23275224, 23245124

E-mail: info@rameshpublishinghouse.com
For Online Shopping: www.rameshpublishinghouse.com
Showroom
● Balaji Market, Nai Sarak, Delhi-6 ✆ 23253720, 23282525
● 4457, Nai Sarak, Delhi-6, ✆ 23918938

Book Code: R-1008

39th Edition: May 2025

ISBN: 978-81-7812-523-7

Price: ₹ 60

Printed at: Deepak Offset, Delhi

CONTENT

THE PERIODIC TABLE

All the naturally occurring elements can be classified in terms of their **atomic structure** and physical or chemical properties. Physical properties include melting and boiling points, atomic volume (relative atomic mass divided by density) and mechanical properties like compressibility (reciprocal of **bulk modulus**). Chemical properties cover such characteristics as oxidation number of **valency**, ionic and **atomic radii** as well as **electronegativity**. These properties (see **Properties of the elements)** all show periodic associations (or *periodicity*) as a function of atomic number. If the elements are arranged in sequence of atomic number starting with hydrogen (Z = 1) from left to right as in the table below, elements in vertical columns (or groups) tend to show similar properties. The first elements hydrogen is ambiguous as it shows both metallic and non-metallic behaviour, forming a cation or an anion depending on its environment. In the Periodic Table are found on the left of the bold line and non-metals on the right of the dotted line. Between metals and non-metals are a cluster of six semi-metals starting with boron which have intermediate properties such as semiconductivity. There are also a group of 14 lanthanides or rare earth elements with very similar properties and four radioactive heavy metals, the actinides.

The Periodic Table with the elements arranged in 18 groups, 1a-0

(See the appropriate entries for the following transuranic elements; neptunium; plutonium; americium; curium; berkelium; califormium; einsteinium; fermium; mendelevium; nobelium; lawrencium; dubnium; joliotium; rutherfordium; bohrium; hahnium; meitnerium.)

1a	2a	3b	4b	5b	6b	7b	8			1b	2b	3a	4a	5a	6a	7a	0
1 H	Metals											Non-metals				1 H	2 He
3 Li	4 Be											5 B	6 C	7 N	8 O	9 F	10 Ne
11 Na	12 Mg											13 Al	14 Si	15 P	16 S	17 Cl	18 Ar
19 K	20 Ca	21 Sc	22 Ti	23 V	24 Cr	25 Mn	26 Fe	27 Co	28 Ni	29 Cu	30 Zn	31 Ga	32 Ge	33 As	34 Se	35 Br	36 Kr
37 Rb	38 Sr	39 Y	40 Zr	41 Nb	42 Mo	43 Tc	44 Ru	45 Rh	46 Pd	47 Ag	48 Cd	49 In	50 Sn	51 Sb	52 Te	53 I	54 Xe
55 Cs	56 Ba	57* to 71	72 Hf	73 Ta	74 W	75 Re	76 Os	77 Ir	78 Pt	79 Au	80 Hg	81 Tl	82 Pb	83 Bi	84 Po	85 At	86 Rn
87 Fr	88 Ra	89⁺ to 92	Metals													Non-metals	
*Lanthanides		57 La	58 Ce	59 Pr	60 Nd	61 Pm	62 Sm	63 Eu	64 Gd	65 Tb	66 Dy	67 Ho	68 Er	69 Tm	70 Yb	71 Lu	
⁺Actinides		89 Ac	90 Th	91 Pa	92 U												

The principal of the elements

The tables on the following three pages list of the all the naturally occurring elements alphabetically together with some of their properties. The following abbreviations are used.

bc = body centred; bcc = body-centred cubic
cubic (diam) = diamond structure
Electroneg = Electronegativities as assigned by Pauling on a scale of 0 to 4
fc = face centred; fcc = face-centred cubic
graph = graphite
hcp = hexagonal close-packed
hex = hexagonal
mon = monoclinic
ortho = orthorhombic
Ox. Nos = principal oxidation numbers
r = red and y = yellow
r_a, r_1 = atomic, ionic radii calculated from closest distances between atoms in crystal structure
r.a.m. = rhombohedral
tetra = tetragonal
T_m, T_b = temperatures of melting and boiling, all exc. arsenic, helium at atmospheric pressure
T_{TRANS} = temperature for transformation in crystal structure column
Z = atomic number

1

INTRODUCTION

Fundamental 7 Units in SI

Base Quantity	Name of Unit	Symbol
Length	meter	m
Mass	kilogram	kg
Time	second	s
Electrical current	Ampere	A
Temperature	Kelvin	K
Amount of Substance	Mole	mol
Luminous intensity	Candela	Cd

Supplementary Units

There are two supplementary units, which are as follows:

Physical quantity	Supplementary Units	
	Name	Symbol
1. Plane angle	radian	rad
2. Solid angle	steradian	sr

Derived Units in SI

Quantity	S.I. Unit	Special Name
Area	m^2	
Volume	m^3	
Density	kg/m^3	
Velocity	m/s	
Acceleration	m/s^2	
Force	kgm/s^2	newton, N
Frequency	cycles/s	hertz, Hz
Energy	$kg\text{-}m^2/s^2$	joule, J
Concentration	$mole/m^3$	
Molar mass	kg/mol	
Pressure	$kg/m\text{-}s^2$	pascal, Pa

SI Prefixes for Multiples and Fractions of SI Units

Multiple	Prefix	Symbol
10^{18}	exa	E
10^{15}	peta	P
10^{12}	tera	T
10^{9}	giga	G
10^{6}	mega	M
10^{3}	kilo	K
10^{2}	hecto	h
10^{1}	deca	da
10^{-1}	deci	d
10^{-2}	centi	c
10^{-3}	milli	m
10^{-6}	micro	μ
10^{-9}	nano	n
10^{-12}	pico	p
10^{-15}	femto	f
10^{-18}	atto	a

Meter → It is defined as the length equal to 1,650,763.73 wavelengths in vacuum of the orange red of the spectrum of krypton-86.

Kilogram → It is defined as the mass of a cylinder of platinum-iridium alloy kept by the international bureau of weights and measures at Paris.

Second → It is defined as the duration of 9,192,631,770 cycles of the radiation associated with specified transition of cesium 133.

Ampere → It is defined as the current that, when flowing through each of two long parallel wires separated by 1 meter of free space, results in a force between the wires of 2×10^{-7} newton per meter of length.

Kelvin → It is defined as the fraction $\frac{1}{273.16}$ of the **temperature of the absolute zero,** triple point of water.

Mole → It is defined as the amount of a substance that contains as many entities as there are atoms in exactly 0.012 kilogram of carbon-12.

Candela → It is defined as the luminous intensity of $\frac{1}{600,000}$ of a square meter of a black body at the temperature of freezing platinum (2045 K).

Do you know?

• Oxalic acid is obtained from	:	Cane sugar and sorrel plant
• Formic acid is obtained from	:	Red ants
• Uric acid is obtained from	:	Urine
• Glycerine is obtained from	:	Olive oil
• Citric acid is obtained from	:	Lemon
• Malic acid is obtained from	:	Apples
• Lactic acid is obtained from	:	Sour milk

2

ATOMIC STRUCTURE

K = °C + 273.15

1 mole = 6.023×10^{23} particles

1 amu = 1.662×10^{-24} g

n = $\frac{m}{M}$; where n = number of moles of an element

m = mass of the element

M = gram atomic mass of the element

Atomic Number (Z) = No. of protons in the nucleus

= no. of electrons in neutral atom.

Mass Number (A) = No. of neutrons + No. of protons, i.e.

A = n + p

No. of neutrons = A – Z.

Maximum no. of electrons in an orbit = $2n^2$

where n = orbit no. or principal quantum number

Maximum no. of electrons in a Sub-shell

$= 4n - 2$

$= 2(2l + 1)$

$= 4l + 2$

where l = azimuthal quantum number.

- For any value of n, there are n values of l and the value ranges from 0 to n – 1.
- For any value of l, there are $2l + 1$ values of m and the value ranges from $-l$ to $+l$ including zero.

  where, m = magnetic quantum number
- For each value of n, there are n^2 values of m.
- For a particular value of n, there are $2n^2$ sets of quantum numbers.

Electron

The cathode ray particles having negative charge and a specific charge to mass ratio remains constant irrespective of the nature of gas in discharge tube are known as electrons.

$\frac{e}{m}$ ratio of an electron is given by,

$$\frac{e}{m} = \frac{E}{B^2 R}$$

Where B = applied magnetic field
R = readius of circular path of electron
E = applied electric field

The value of $\frac{e}{m}$ was calculated as 1.77×10^{11} C kg^{-1}

Charge of electron
$= -1.59 \times 10^{-19}$ C $= -1.6 \times 10^{-19}$ C
So that, mass of electron $= 9.107 \times 10^{-31}$ kg

Proton

The fundamental particle of the anode ray particles having positive charge and also which is an integral multiple of electronic charge and a specific charge to mass ratio, both depending on the nature of the gas in discharge tube are known as protons.

The value of $\frac{e}{m}$ of proton was found to be

9.58×10^{7} C kg^{-1}
Charge on proton $= +1.6 \times 10^{-19}$ C
So that, mass of proton $= 1.67 \times 10^{-27}$ kg

Neutron

A neutral particle having nearly the same mass as that of the proton, discovered by Sir James Chadwick is known as the neutron.

Thomson's model of atom

An atom is a sphere of positive charges with negative charges embedded in it, such that the atom as a whole remains neutral.

This model is also known as *water-melon model of atom*. But this model could not explain the Rutherford's gold-foil scattering experiment.

Rutherford's model of atom

I. An atom consists of a positively charged heavy central core, called as *nucleus*.

II. The size of the nucleus is very small as compared to the total size of the atom.

III. The electrons revolve round the nucleus and their total negative charge is equal to the positive charge of the nucleus, so that the atom as a whole is electrically netural so they form the extra-nuclear region of the atom.

Limitations of Rutherford's theory:

I. If the electrons were at rest round the nucleus, they will be attracted by it and will finally fall into it.

II. If the electrons were revolving round the nucleus in circular orbits, they will radiate out energy continuously and hence spirally fall into the nucleus.

Isotopes

Atoms of same element are said to be isotopes of that element if they have same atomic number but different mass numbers or in other words, same number of protons and electrons but different neutron number.

such as $^{1}_{1}H$, $^{2}_{1}H$, $^{3}_{1}H$; $^{12}_{6}C$, $^{13}_{6}C$, $^{14}_{6}C$; $^{16}_{8}O$, $^{17}_{8}O$, $^{18}_{8}O$,

Isobars

Atoms of different elements are said to be isobars if they have same mass numbers though their atomic numbers are different. i.e., inspite of different number of neutrons and protons, the sum of $n + p$ is the same.

such as $^{40}_{18}Ar$, $^{40}_{19}K$, $^{40}_{20}Ca$.

Isotones

Atoms of different elements are said to be isotones if they have the same number of neutrons even though the atomic numbers and mass numbers are different.

Bohr's model of atom

Postulates of Bohr's theory

(*i*) The electrons revolve round the nucleus in closed circular orbits. Necessary centripetal forces for rotation is provided by the electrostatic force between the nucleus and the electon.

(*ii*) An electon can revolve only in certain discrete, non-radiating orbits, called stationary orbits. For these orbits, the total angular momentum of the moving electron is an integral multiple of $\frac{h}{2\pi}$.

(*iii*) The energy is radiated only when an electron jumps from one stationary orbit to another.

Atomic quantum numbers

1. **Principal quantum number (*n*):** The principal quantum number n corresponds to the principal energy level of the elctrons.

It has integral positive values excluding zero.

2. **Angular quantum number or azimuthal quantum number (l):** Having the same principal energy level, there are different orbitals having different energy and different angular momentum. These orbitals are called *subshells* and are characterized by their orbital quantum number l, l can have all integral values from 0 to $(n - 1)$. If $l = 0, 1, 2, 3$ ... the sub-shells are called as *s*, *p*, *d*, *f* ..... respectively.

Shells	Values of n	Values of l	Representation of sub-shells
K	1	0	1s
L	2	0, 1	2s, 2p
M	3	0, 1, 2	3s, 3p, 3d
N	4	0, 1, 2, 3	4s, 4p, 4d, 4f

3. **Magnetic quantum number (m):** The changing the orientation of the electrons in space around the nucleus are determined by magnetic quantum numbers m. For a given value l, m can have any integral value from $-l$ to $+l$ including zero, i.e., for a given value of l, m can have of $(2l + 1)$ values.

Sub-Shells	Values of l	Values of m	Number of orbitals
s	0	0	1
p	1	−1, 0, +1	3
d	2	−2, −1, 0, +1, +2	5
f	3	−3, −2, −1, 0, +1, +2, +3	7

4. **Spin quantum number (s):** Spin quantum number determines the spining of an electron about its own axis and can have only two directions: clockwise or anti clockwise, can have either of the two values, $+\frac{1}{2}$ or $-\frac{1}{2}$.

Pauli's Exclusion Principle

It states that no two electrons in an atom can have the same values of all the four quantum numbers. In other words, no two electrons can occupy the same energy state.

$(n + l)$ rule

It states that a sub-shell having lower value of $(n + l)$ has lower energy. If the value of $(n + l)$ for two sub-shells is the same, then the one having lower value of n has lower energy.

Aufbau principle

It states that electrons enter the various orbitals of an atom according to their increasing energies, and energy of the orbitals is governed by (n + l) rule. The electrons enter sub-shell of lowest energy first.

Hund's rules

Rule I. When there are several orbitals of equal energy available, first of all electrons enter these orbitals singly, pairing of electrons takes place later.

Rule II. The spins of all these electrons in orbitals of equal energies, should be parallel.

Bond order : It is a measure of the strength of the chemical bond. It is equal to the number of covalent bonds in a molecule. Now,

Bond order

$$= \frac{\begin{bmatrix}\text{Number of electrons in} \\ \text{bonding molecular orbitals}\end{bmatrix} - \begin{bmatrix}\text{Number of electrons in} \\ \text{antibonding molecular orbitals}\end{bmatrix}}{2}$$

$= (N_B - N_A)/2$

(*i*) Greater the bond order, greater is the stability of the molecule.

(*ii*) Bond length is inversely proportional to the bond order.

(*iii*) Bond dissociation energy is directly proportional to the bond order.

Hybridization

It is used to explain the shapes of molecules. *It is a process of inter mixing of orbitals of slightly different energies and redistribute their energies to give rise to new set of orbitals of same energy, size and shape.*

The new orbitals formed are called **hybrid** or **hybridized orbitals.** The number of hybridized orbitals formed is equal to the number of combining pure orbitals.

Energy of Electron in nth Orbit (E_n)

(a) In C.G.S. unit :

$$E_n = \frac{2\pi^2 m z^2 e^4}{n^2 h^2} \text{ erg / electron}$$

In another way, $E_n = -\frac{E_1}{n^2}$

where, m = mass of electron = 9.11×10^{-28} gm

e = charge of electron = -4.8×10^{-10} esu

Types of hybridization

Type of hybridization	Atomic orbitals	Bond angle	Orientation of hybridized orbitals	Examples
sp	$s + p$	180°	Linear	$BeCl_2$, $HgCl_2$, $ZnCl_2$, N_2O, C_2H_2,
sp^2	$s + 2(p)$	120°	Triangular planar	BCl_3, BF_3, $AlCl_3$, SO_2, SO_3, C_2H_4, CO_2^{-2}, NO_3^-
sp^3	$s + 3(p)$	109° 28'	Tetrahedral	$SiCl_4$, SiF_4, $SnCl_4$, CCl_4, CH_4, NH_4^+, BF_4^-
dsp^2	$d + s + 2(p)$	90°	Square planar	$[PtCl_4]^{2-}$, $[Ni(CN)_4]^{2-}$ $[Cu(NH_3)_4]^{2+}$
dsp^3 or sp^3d	$d + s + 3(p)$	90°, 120°	Trigonal bipyramidal	PF_5, PCl_5, SbF_5
d^2sp^3, or sp^3d^2	$2(d) + s + 3(p)$	90°	Octahedral	SF_6, $[SiF_6]^{2-}$, CrF_6^{3-}, $[Co(CN)_6]^{3-}$, $[Co(NH_3)_6]^{3+}$
d^3sp^3 or sp^3d^3	$3(d) + s + 3(p)$	90°, 72°	Pentagonal bipyramidal	IF_7

z = atomic number

h = Planck's constant = 6.625×10^{-27} erg-sec

E_n and E_1 represents the energies of electron in nth and 1st orbit respectively.

(b) In S.I. unit :

$$E_n = -\frac{2\pi^2 m z^2 e^4}{(4\pi \epsilon_0)^2 n^2 h^2}$$

$$= -\frac{m z^2 e^4}{8 \epsilon_0^2 n^2 h^2} \text{ joule / electron}$$

where, m = mass of electron = 9.11×10^{-31} kg

e = charge of an electron

= -1.6×10^{-19} coulomb

z = atomic number

h = Planck's constant = 6.625×10^{-34} joule-sec

$4\pi\epsilon_0$ is called permittivity factor and is equal to

$$\frac{1}{9 \times 10^9} \frac{(\text{Coulomb})^2}{\text{Newton} \times (\text{metre})^2} = \frac{1}{9 \times 10^9} \frac{C^2}{N \times m^2};$$

ϵ_0 is called permittivity of free space and is equal to

$$\frac{1}{36\pi \times 10^9} \frac{C^2}{N \times m^2} = 8.85 \times 10^{-12} \frac{C^2}{N \times m^2}$$

Speed of the nth Stationary Orbit (v_n)

(a) In C.G.S. unit

$$v_n = \frac{2\pi ze^2}{nh} \text{ cm / sec}$$

In another way

$$v_n = \frac{v_1}{n}$$

where v_n and v_1 represent the velocities of electron in nth and 1st orbit respectively.

(b) In S.I. unit

$$v_n = \frac{ze^2}{2 \epsilon_0 nh} \text{ m / sec}$$

Angular momentum of electron in nth orbit :

$$I\omega_n = mv_n r_n = \frac{nh}{2\pi}$$

- Radius of atom $\approx 10^{-8}$ cm = 1 Angstrom unit (1 Å)
- Radius of nucleus $\approx 10^{-13}$cm = 1 Fermi

E = h.ν, where E = energy; ν = Frequency of radiation

- Wt. of a single particle (atom, molecule, ion, etc.) = $\frac{\text{gram formula wt.}}{N}$
- Molar volume of gas = 22.4 litres at S.T.P. or N.T.P.
- Normal temperature = 0°C or 273°K
- Normal pressure = 1 atmosphere or 700 mm of Hg
- Vapour density (V.D.) = $\frac{W}{V} \times 11.2$
- Molecular Wt. (M) = $\frac{W}{V} \times 22.4$; where

  V is the volume in litres at S.T.P. or N.T.P.
- M = 2 × V.D.
- No. of moles after decomposition or dissociation

= 1 + (n – 1) × α; where α is degree of dissociation and n = no. of particles after dissociation

- $\frac{M_c}{M_o} = \frac{D_c}{D_o} = 1 + (n - 1) \times \alpha$ in case of dissociation
- $\frac{M_c}{M_o} = \frac{D_c}{D_o} = 1 - \alpha\left(\frac{n-1}{n}\right)$ in case of association

  where α = degree of association, c stands for calculated or normal value and o stands for observed or abnormal value.
- Equivalent weight of an element

  $$= \frac{\text{At. wt.}}{\text{Valency}}$$
- 1 gm eq. wt. = eq. wt. in gram.

  No. of gram equivalent

  $$= \frac{\text{Wt. in gram}}{\text{gram equivalent wt.}}$$
- Equivalent wt. of an acid

  $$= \frac{\text{Molecular wt.}}{\text{Basicity of an acid}}$$

- Equivalent wt. of base

$$= \frac{\text{Molecular wt.}}{\text{Acidity of a base}}$$

- If no redox reaction, then equivalent wt. of a radical

$$= \frac{\text{Formula wt.}}{\text{Charge on ions (radicals) or valency}}$$

- Eq. wt. of a compound

$$= \frac{\text{Molecular wt.}}{\text{Total charge of its cation or anion}}$$

= sum of equal wt. elements and radicals
= Eq. wt. of cation + Eq. wt. of anion

- For reaction with H_2 : $E = \frac{W_1}{W_2(H)}$

$$= \frac{\text{wt. of metal}}{\text{wt. of } H_2}$$

$$= \frac{W_1}{V} \times 11{,}200$$

- For reaction with chlorine :

$$E = \frac{W_1}{W_2(Cl)} \times 35.5$$

$$= \frac{W}{V} \times 11{,}200$$

$$= \frac{\text{wt. of element}}{\text{wt. of chlorine}} \times 35.5$$

- For reaction with oxygen :

$$E = \frac{W_1}{W_2(O_2)} \times 8$$

$$= \frac{\text{wt. of substance}}{\text{wt. of Oxygen}} \times 8$$

$$= \frac{W_1}{V} \times 5{,}600$$

In general reaction : $\dfrac{W_1}{W_2} = \dfrac{E_1}{E_2}$

In redox reaction : Equivalent wt. of a

$$\text{substance} = \frac{\text{Formula wt.}}{\text{No. of electrons lost or gained per molecule}}$$

$$= \frac{\text{Formula wt.}}{\text{Total change in oxidation number per molecule}}$$

For an isomorphous AX, BX

$$\frac{\text{wt. of A that combines with x gm. of X}}{\text{wt. of B that combines with x gm. of X}}$$

$$= \frac{\text{At. wt. of A}}{\text{At. wt. of B}}$$

3

PERIODIC TABLE

1. **The periodic table:** Earlier concepts of classifying elements

   (*i*) **Doebereinier's law of Triads:** Doebereinier classified the elements into the groups of three elements, each was called as 'triads.' The atomic weight of the middle element in each triad was found to be the average of the atomic weights of the two neighbouring elements. *e.g.*

Triad	I	II	III
Elements	Li	Na	K
At. wt.	7	23	39

At. wt. of the elements (Na)

$$= \frac{7+39}{2} = 23$$

This classification failed, as it could not classify all the known elements.

(*ii*) **Newlands' law of octaves :** Newlands arranged the elements in the increasing order of atomic weights and found that the properties of eighth element resemble closely with the starting element. *e.g.*

Elements	Li	Be	B	C	N	O	F
At. wt.	7	9	11	12	14	16	19

This law could not be applied for heavier elements.

(*iii*) **Mendeleef's periodic law and periodic table :** In 1869 Mendeleef arranged 63 elements into a periodic table on the basis of his periodic law, which states that *physical and chemical properties of the elements are periodic function of their atomic weights.*

This table has 8 vertical columns known as groups and 7 horizontal rows known as periods. Properties of elements are repeated after every seventh elements. Elements in the same group show similar properties.

Merits of Mendelef's Periodic Table

I. Mendeleff predicted the existance of certain elements and left spaces in his periodic table, these elements were later on discovered.

II. Mendeleef correct the atomic weights of certain elements, such as Au and Pt.

III. Mendeleef corrected the valence of certain elements, like Be.

Demerits of Mendeleef's Periodic Table

I. Position of hydrogen was not clear.

II. No distinction was made between metals and non-metals.

III. No place was assigned to the isotopes of an element.

IV. Variable valances of the elements could not be explained.

V. Certain dissimilar elements were placed in the same group, while certain similar elements were placed in different groups.

VI. Some elements with higher atomic weights were placed before the elements with lower atomic weights.

VII. Elements of group VIII were placed in a period.

2. The modern periodic table

Modern periodic law : It states that physical and chemical properties of the elements are periodic function of their atomic numbers. Hence,

(*i*) The elements were arranged in order of their increasing atomic numbers.

(*ii*) The elements resembling in their physical and chemical properties fall one below another.

(*iii*) There are 7 horizontal rows called as **periods** and 18 vertical columns called as **groups**.

Periods

Horizontal rows are known as periods. There are 7 periods. The number of period represents the number of orbits. Each period begins with an alkali metal and ends up with a noble gas. However, first period begin with hydrogen.

(*i*) The first period contains 2 elements, H_1 and He_2.

(*ii*) The second and third periods contain 8 elements each and are called as short periods. The elements are from lithium (Li_3)

to neon (Ne_{10}) and from sodium (Na_{11}) to argon (Ar_{18}).

(*iii*) The fourth and fifth periods contain 18 elements each and are called as long periods. The elements are from potassium (K_{19}) to krypton (Kr_{36}) and from rubedium (Rb_{37}) to xenon (Xe_{54}).

(*iv*) Sixth period contains 32 elements and is called as very long period. The elements contained are from caesium (Cs_{55}) to radon (Rn_{86}).

(*v*) Seventh period contains 21 elements, from francium (F_{87}) to unnitseptium (Uns_{107}).

(*vi*) The 14 elements each and lanthanum and actinium belonging to 6th and 7th periods are placed in two separate rows at the bottom known as **lanthanids** (57-71) and **actinides** (89-103).

Groups

18 vertical columns are known as groups. Elements in a group have similar physical and chemical properties because of similar outer shell electronic configuration.

(*i*) Elements of groups 1, 2, 13-18 are called as *normal, typical or representative elements.* They have their innermost orbit complete (except group 18) and outer electronic configurations as ns^1, ns^2, ns^2np^1, ns^2np^2, ns^2np^3, ns^2np^4, ns^2np^5 and ns^2np^6 repectively. Group 1 is known as *alkali metals group,* group 2 is called *alkaline earth metals group* and group 17 is known as *halogen group.* The elements of group 18 are *noble gases.*

(*ii*) Elements of group 3-12 are called as *transition elements* as their properties lie in between those of metals and non-metals. They have their outermost as well as penultimate shells incomplete. Their general electronic configuration is $(n-1)\,d^{1-10}ns^{0-2}$.

(*iii*) Inner transition metals are placed in two separate horizontal rows below seventh period known as Lanthanons and actinons.

3. **Electronic structure and the periodic table**

(*i*) **s-block elements:** Those elements in which the last electron enters into the outermost s-subshell are called a s-block elements. These are the elements of group 1 having outermost electronic configuration ns^1 and elements of group 2 having outermost electronic configurations ns^2.

(ii) **p-block elements:** Those elements in which the last electron enters into the p-orbitals of outermost shell are called as p-block elements. These are the elements of groups 13-18 (excluding He). The general outer electronic configuration varies from ns^2np^1 to ns^2np^2.

(iii) **d-block elements:** Those elements in which the last electron enters in (n–1) d orbitals are called as d-block elements. This block includes elements of groups 3-12. This block lies between s-block and p-block and hence its elements are called as transition elements. They have configuration $(n-1)\, s^2\, p^6\, d^{1-10}, ns^{1-2}$.

(iv) **f-block elements:** Those elements in which the last electron enters in (n–2) f

orbitals are called as f-block elements. There are two series of f-block elements: *Lanthanides* having incomplete 4f-orbitals [from cerium (58) to lutetium (71)] and *Actinides* having incomplete 5 f-orbitals [from thorium (90) to lawrencium (103)].

4. **Periodic properties of the elements:** The cause of periodic properties of the elements is due to regular repetition of similar outer shell electronic configuration after each period.

   Main periodic properties of elements are :

   (*i*) Electronic configuration and valence

   In periods: The elements in period have different electronic configurations and different number of valence electrons.

   In groups: The elements of same group have similar outer shell electronic configuration and hence same valence.

   (*ii*) Atomic size: The distance from the atom to the outermost shell of the electrons is called atomic radius.

In periods: On moving from left to right in a period, atomic radius decreases.

In groups: Atomic size increases as we move down in a group.

(*iii*) Ionization energy: Minimum energy needed to remove the most loosely bound electron from an isolated gaseous atom or cation, is called as ionization energy.

In periods: The ionization energy of the elements increases as we move from left to right.

In groups : On moving down in a group, ionization energy decreases.

(*iv*) Electron affinity: The energy released when an electron is added to a neutral isolated gaseous atom is called electron affinity.

In periods: Electron affinity increases as we move across the period from left to right. Halogens have the highest electron affinity. Electron affinity of rare gases, half filled and completely filled orbitals is zero (due to extra stability).

(*v*) **Electronegativity** : Power of an elements to attract shared pair of electrons of a covalent bond in a compound towards itself, is called its electronegativity.

In periods: It increases along a period from left to right.

In groups: it decreases down the groups.

4

CHEMICAL BONDING

Noble Gases-Octet Rule

Noble gases are also known as inert gases or rare gases. These gases are stable and exist in free state because they have their outermost orbit completely filled. Hence, they are chemically unreactive or intert.

Electronic configurations of noble gases

Element	Symbol	Atomic number	Electronic configuration	Electronic structure
Helium	He	2	2	$1s^2$
Neon	Ne	10	2,8	$1s^2 2s^2 2p^6$
Argon	Ar	18	2, 8, 8	$1s^2 2s^2 2p^6 3s^2 3p^6$
Krypton	Kr	36	2, 8, 18, 8	$1s^2 2s^2 2p^6 3s^2 3p^6 3d^{10} 4s^2 4p^6$
Xenon	Xe	54	2, 8, 18, 18, 8	$1s^2 2s^2 2p^6 3s^2 3p^6 3d^{10} 4s^2 4p^6 4d^{10} 5s^2 5p^6$
Radon	Rn	86	2, 8, 18, 32, 18, 8	$1s^2 2s^2 2p^6 3s^2 3p^6 3d^{10} 4s^2 4p^6 4d^{10} 5s^2 5p^6 5d^{10} 6s^2 6p^6$

All noble gases (except Helium) have eight electrons in their outermost shell. This led to idea of octet rule which states that *atom tends to lose, gain or share valence electrons with other atom to achieve the state of noble gases.*

Causes of Chemical Combination

The cause of chemical combinatioin is to gain the state of noble gases and thus the state of lower energy, extra stability, low electron affinity, high ionisation energy and chemical inertness. This can be done:

(a) By loss or gain of valence electrons:

Sodium atom loses one elctron to attain the configuration of neon atom:

$$Na \rightarrow Na^+ + e^-$$

sodium atom (2, 8, 1)	Sodium ion (2, 8)	Electron

Chlorine atom gains one electron to attain the configuration of argon atom:

$$Cl + e^- \rightarrow Cl^-$$

Chlorine atom (2, 8, 7)	Electron	Chloride ion (2, 8, 8)

Hence,

$$Na^+ + Cl^- \rightarrow NaCl$$

Sodium ion	Chloride ion	Sodium chloride

(b) By sharing of valence electrons with other atom:

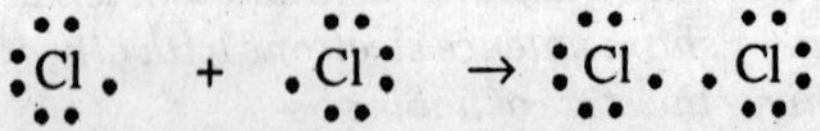

$:\ddot{\underset{\cdot\cdot}{Cl}}\cdot \;+\; \cdot\ddot{\underset{\cdot\cdot}{Cl}}: \;\rightarrow\; :\ddot{\underset{\cdot\cdot}{Cl}}\cdot\,\cdot\ddot{\underset{\cdot\cdot}{Cl}}:$

Chlorine atom   Chlorine atom   Chlorine molecule

During the formation of chemical bond, energy is released and so the resultant molecule has lower energy and greater stability.

1. Ionic or electrovalent bonding:

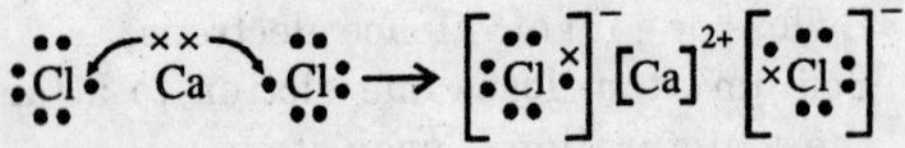

$:\ddot{\underset{\cdot\cdot}{Cl}}\cdot \;\overset{\times\times}{Ca}\; \cdot\ddot{\underset{\cdot\cdot}{Cl}}: \rightarrow [:\ddot{\underset{\cdot\cdot}{Cl}}\overset{\times}{\cdot}]^{-}\;[Ca]^{2+}\;[\overset{\cdot}{\times}\ddot{\underset{\cdot\cdot}{Cl}}:]^{-}$

Chlorine atom (2, 8, 7)   Calcium atom (2, 8, 8, 2)   Chlorine atom (2, 8, 7)   Chloride ion (2, 8, 8)   Calcium ion (2, 8, 8)   Chloride ion (2, 8, 8)

2. Covalent bonding:

(*i*) $\cdot\ddot{\underset{\cdot\cdot}{O}}: + \overset{\times}{\underset{\times}{}}\overset{\times\times}{\underset{\times\times}{O}} \rightarrow$ Oxygen molecule or $O = O$

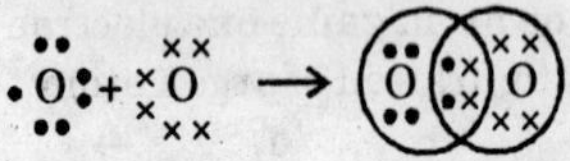

Oxygen atom   Oxygen atom   Oxygen molecule

(*ii*) $\ddot{N}: + \overset{\times\times}{N}$ $\rightarrow$ Nitrogen molecule or $N \equiv N$

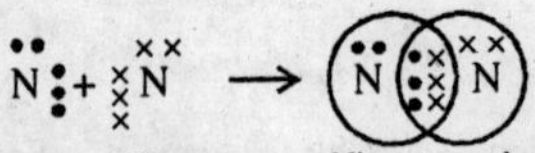

Nitrogen atom   Nitrogen atom   Nitrogen molecule

(*iii*) $\overset{\times}{H} + \cdot\ddot{\underset{\cdot\cdot}{Cl}}: \longrightarrow$ H(Cl) or H – Cl

Hydrogen atom   Chlorine atom   Hydrogen chloride molecule

Co-ordinate Covalent Bond

In this bond shared pair of electrons (lone-pair) comes from one atom only. The atom which donates lone pair is known as donor and other atom which accepts, it is known as acceptor. It is represented by an arrow (→) from doner to acceptor atom.

(*a*) Formation of ammonium ion

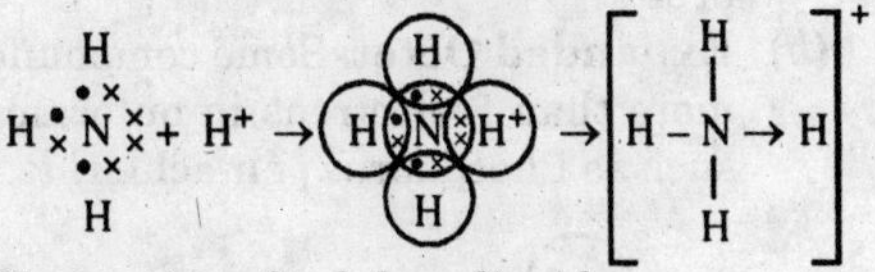

$H_3N + H^+ \rightarrow [NH_4]^+$

(*b*) Formation of sulphur dioxide

$O + S + O \rightarrow O = S \rightarrow O$

Deviation from Octet Rule

In some cases, it was observed that atoms do not follow octet rule during bonding. Instead of having followed octet rule, the molecules have,

(*a*) odd number of electrons.
(*b*) expanded octet.
(*c*) incomplete octet.

(*a*) **Odd number of electrons:** Normally molecules have even number of electrons and thus complete pair of electrons. But some molecules like ClO_2, NO and NO_2 have odd number of electrons. Such as nitric oxide

N has 7 electrons in outermost orbit.

(*b*) **Expanded Octet:** Some compounds have more than 8 electrons in outermost shell such as Phosphorus pentachloride.

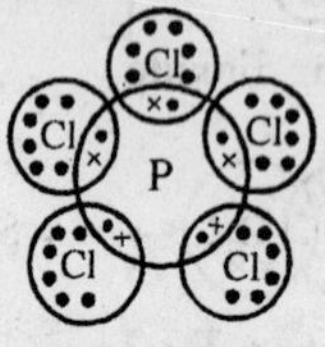

P has 10 electrons in its valence shell.

(*c*) **Imcomplete octet:** Some compounds are fewer number of electrons than 8. Such as Boron trifluoride.

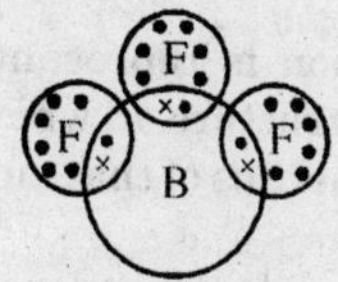

BF_3 has 6 electrons in outermost shell of B atom.

3. Polar covalent bond:

$$H^{\bullet} + {}_{\times}\overset{\times\times}{\underset{\times\times}{Cl}}\,\overset{\times}{\underset{\times}{}} \longrightarrow H\overset{\bullet}{\times}\overset{\times\times}{\underset{\times\times}{Cl}}\,\overset{\times}{\underset{\times}{}} \text{ or } \overset{\delta+}{H}-\overset{\delta-}{Cl}$$

4. Coordinate covalent bond:

$$H\overset{\bullet}{\times}\overset{\overset{H}{\bullet\times}}{\underset{\underset{H}{\bullet\times}}{N}}\overset{\times}{\underset{\times}{}} + H^{+} \longrightarrow \left[\begin{array}{c} H \\ | \\ H - N \rightarrow H \\ | \\ H \end{array}\right]^{+}$$

5. Electropositive element + Electronegative element → Ionic bond

   Electronegative element + Electronegative element → Covalent bond

   Electropositive element + Electropositive element → Metallic bond

Resonance

Sometimes molecule or ion is represented by more than one electronic structure, in which only one represents all the properties of that molecule or ion.

Various dot structures for a molecule are known as resonating structures. Actual structure is intermediate of all these resonating structures *e.g.*:

Resonance forms of carbonate ion, CO_3^{2-}

$$:\ddot{O}=C\langle\ ^{\ddot{O}:^{\ominus}}_{\ddot{O}:^{\ominus}} \longleftrightarrow {}^{\ominus}:\ddot{O}-C\langle\ ^{\ddot{O}:^{\ominus}}_{\ddot{O}} \longleftrightarrow {}^{\ominus}:\ddot{O}-C\langle\ ^{\ddot{O}}_{\ddot{O}:^{\ominus}}$$

Sigma (σ) and pi (π) Bonding

Sigma (σ) **bond** between two atoms involves head on overlapping along their internuclear axis which gives rise to maximum electron density on the axis. Overlapping takes place in the following ways:

(*i*) **s-s overlapping:** This type of overlapping takes place between half filled s-orbitals of the atoms, *e.g.* in H_2 molecule.

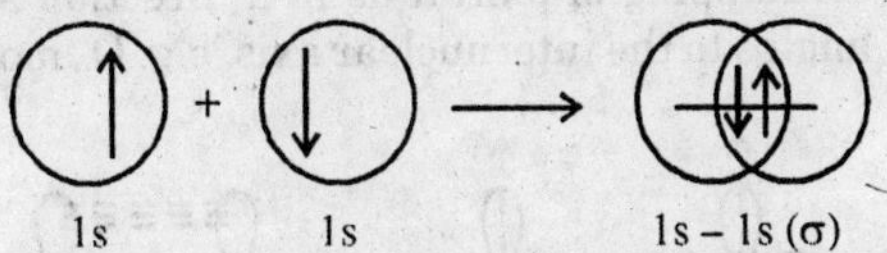

Formation of hydrogen molecule

(*ii*) s-p overlapping: This involves overlapping between half filled s-orbital of one atom with half filled p-orbital of another atom. *e.g.* In HF molecule.

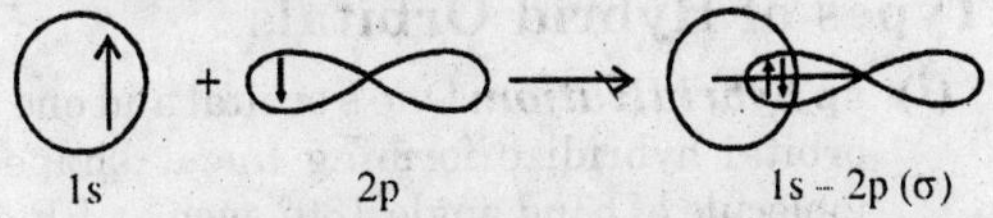

Formation of hydrogen fluoride molecule

(*iii*) p-p overlapping: This involves end to end overlapping of p-orbitals of different atoms. e.g. in F_2 molecule.

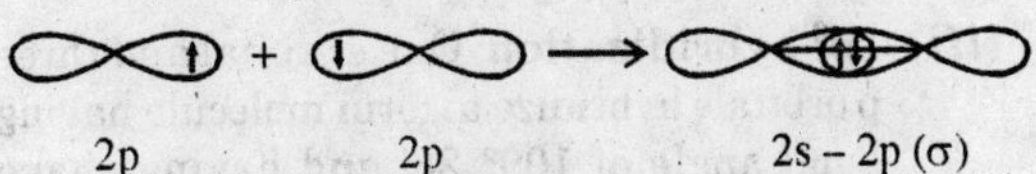

Formation of fluorine molecule

Pi (π) bonding: pi bond is formed by lateral or side ways overlapping of p-orbitals i.e. by overlapping of p-orbitals in a direction at right angles to the internuclear axis. e.g. O_2 molecule.

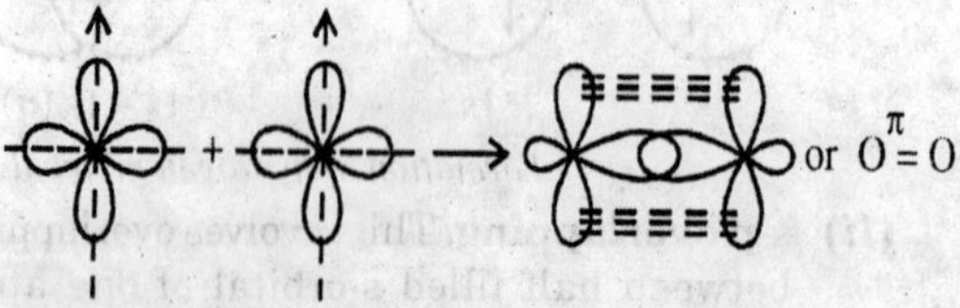

Two half filled p-orbitrals of O-atom *Oxygen molecule*

Types of Hybrid Orbitals

(*i*) **sp *hybridization*:** One s orbital and one p orbital hybridize forming linear shaped molecule of bond angle 180° such as BF_2.

(*ii*) **sp^2 hybridization:** One s orbital and two p orbitals hybridize to form molecule planar triangular in shape with bond angle of 120°. such as C_2H_2, N_2O, BCl_3, SO_3.

(*iii*) **sp^3 hybridization:** One s orbital and three p orbitals hybridize to form molecule having bond angle of 109° 28′ and having shape tetrahedral.

(*iv*) **sp^3 d hybridization:** One s, three p and one d orbitals hybridize to give sp^{3}d hybridization. It has trigonal bipyramidal shape. Such as PCl_5, SF_4, XeF_2, ICl_3, ClF_3.

(*v*) **sp^3d^2 hybridization:** One s, 3p and 2d orbitals are hybridized. Shape of the molecule is octahedral. It has 6 hybridized orbitals. e.g. SF_6.

(*vi*) **sp^3d^3 hybridization:** One s, 3p and 3d orbitals are hybridized to give 7 hybridized orbitals. Five are coplanar and two are perpendicular to this plane. e.g. Cl_7I.

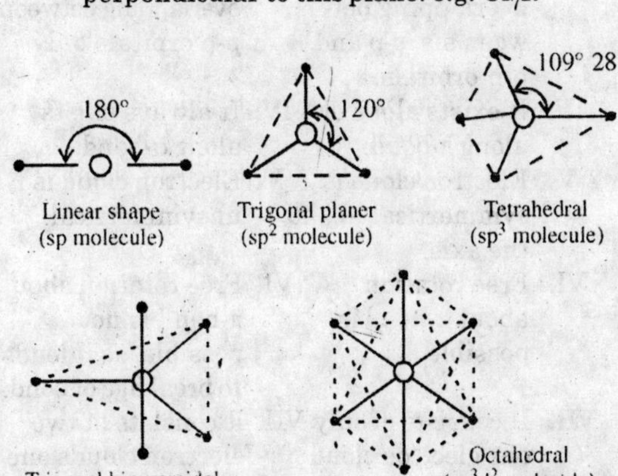

Linear shape (sp molecule)

Trigonal planer (sp^2 molecule)

Tetrahedral (sp^3 molecule)

Trigonal bipyramidal (sp^3d molecule)

Octahedral (sp^3d^2 molecule)

Difference Between Sigma and pi-Bonds

	Sigma (σ) Bond		Pi (π) Bond
I.	It is formed by end to end overlapping of orbitals.	I.	It is formed by side ways overlapping of orbitals.
II.	σ-bond is stronger	II.	π-bond is weaker.
III.	This is formed by overlapping between s-s, s-p and p-p orbitals.	II.	This is formed by overlapping between p-p orbitals only.
IV.	It exists alone or along π-bond.	IV.	It always exists along σ-bond.
V.	Electron cloud is symmertical about the axis.	V.	Electron cloud is unsymmertical.
VI.	Free rotation about σ-bond is possible.	VI.	Free rotation about π-bond is not possible, as it leads to breakage of bond.
VII.	It consists of only one electron cloud about the intern-uclear axis.	VII.	it consists of two electron clouds one above and other below the plane.

Some Common Bond Length

Bond	Bond Length (Å)
C–C (Alkane)	1.54
C=C (Alkene)	1.34
C≡C (Alkyne)	1.20
C–H (Alkane)	1.09
C=C (Benzene)	1.39
O–H (Alcohols)	0.96
C–O (Alcohols)	1.43
C=O (Ketones)	1.21
C–Cl (Chloroalkane)	1.77
N–H (Amines)	1.02
C–N (Amines)	1.47
C=N (Isocyanides)	1.52
C≡N (Cyanides)	1.15
C–Br	1.91
C–I	2.12
H–H	0.74
N≡N	1.09

Bond	Bond Length (Å)
O=O	1.21
F–F	1.42
Cl–Cl	1.99
Br–Br	2.28
I–I	2.67
H–F	0.92
H–Cl	1.27
H–Br	1.41
H–I	1.61

5

STATES OF MATER

Matter : Anything that occupies, space and has some mass is known as matter. There are three physical states of matter.

(*i*) Gaseous state

(*ii*) Liquid state

(*iii*) Solid state

Gas Laws

(*i*) Boyle's law : For given mass of a gas

$$V \propto \frac{1}{P} \text{ or } PV = \text{const.}$$

or $P_1V_1 = P_2V_2$; T constant.

(*ii*) Charles' Law : For given mass of a gas and at constant P.

$$V \propto T \text{ or } \frac{V}{T} = \text{const.}$$

or $\frac{V_1}{T_1} = \frac{V_2}{T_2}$

***(iii)* Gay Lussac's Law** : For given mass of a gas and at constant V,

$$P \propto T \text{ or } \frac{P}{T} = \text{const.}$$

$$\text{or } \frac{P_1}{T_1} = \frac{P_2}{T_2}$$

- t°C = (t + 273)° Abs. or (t + 273)°K
- t°C = T°K – 273 = Abs. – 273
- $V \propto n$, where : P and T are const. (Avagadro's law)

Thus for the same P, V and T, $n_1 = n_2 = n_3 = ...$

Equation of state or Ideal gas equation

(*i*) PV = RT for 1 gm mole

(*ii*) PV = nRT for n gm moles or

$$PV = \frac{m(\text{mass})}{M(\text{mol. wt.})} \times RT$$

where R = Gas const. = 0.0821 litres atm $deg^{-1} mol^{-1}$

= 8.31×10^7 ergs $deg^{-1} mole^{-1}$

= 8.31 joule $deg^{-1} mole^{-1}$

$\simeq$ 2 calories deg^{-1} $mole^{-1}$
= (1.987 calories exactly)

- For the same mass of a gas

$$\frac{P_1V_1}{T_1} = \frac{P_2V_2}{T_2}$$

- If mass changes, number of moles also changes in this case,

$$\frac{P_1V_1}{n_1T_1} = \frac{P_2V_2}{n_2T_2}$$

- $\frac{P_1}{d_1T_1} = \frac{P_2}{d_2T_2}$; where d_1 and d_2 stand for densities.
- Rate of diffusion or effusion (r)

$$= \frac{V\ (\text{volume})}{t\ (\text{time})}$$

- $\frac{r_1}{r_2} = \frac{V_1}{t_1} \Big/ \frac{V_2}{t_2} = \sqrt{\frac{D_2}{D_1}} = \sqrt{\frac{M_2}{M_1}}$

- Partial pressure of a gas (P) = $\frac{nRT}{V}$

where V = volume occupied by the mixture

T = temp. of the mixture and

n = no. moles of the gas.

- Total pressure (P) $= P_1 + P_2 + P_3 + ... + P_n$

$$= \sum n \frac{RT}{V}$$

where $\sum n$ is the total number of moles in the mixture.

- $$\frac{\text{Partial pressure}}{\text{Total pressure}} = \frac{n}{\sum n} = \text{mole fraction}$$

- $$P = \frac{P_1V_1 + P_2V_2 + P_3V_3 + ...}{\text{Total volume}}$$

Symbol Used

P = pressure of gas

C = R.M.S. velocity

V_{av} = mean or average velocity

α = most probable speed

n = no. of molecules of the gas

m = mass of each molecule of the gas

M = mol. wt.

V = volume of one mole of the gas

T = absolute temp.

R = gas const.

λ = mean free path

d = diameter of each molecule

ρ = density of the gas

η = co-efficient of viscosity

K = Boltzman's constant = $\frac{R}{N}$

[N = Avogadro's no.]

$V_{av} = 0.9213 \times C$

$\alpha = 0.816 \times C$

$$PV = \frac{1}{3}mnC^2 = \frac{1}{3}MC^2$$

$$C = \sqrt{\frac{3PV}{M}} = \sqrt{\frac{3P}{\rho}} = \sqrt{\frac{3RT}{M}} = \sqrt{\frac{3KT}{m}}$$

$$V_{av} = \frac{V_1 + V_2 + V_3 + \dots + V_n}{n}$$

Crystal System	Lengths of edges	Interfacial angles	Lattice types	Examples
I. Cubic	a = b = c	$\alpha = \beta = \gamma = 90°$	S*, BC, FC	NaCl, KCl, Zinc blende
II. Tetragonal	a = b ≠ c	$\alpha = \beta = \gamma = 90°$	S, BC	SnO_2, TiO_2, ZnO_2
III. Orthorhombic	a ≠ b ≠ c	$\alpha = \beta = \gamma = 90°$	S, BC, FC, EC	$BaSO_4$, KNO_3, $PbCO_3$
IV. Monoclinic	a ≠ b ≠ c	$\alpha = \gamma = 90° \neq \beta$	S, EC	$CaSO_4 . 2H_2O$, $KClO_3$
V. Triclinic	a ≠ b ≠ c	$\alpha \neq \beta \neq \gamma \neq 90°$	S	$CuSO_4 . 5H_2O$. $K_2Cr_2O_7$
VI. Rhombohedral or Trigonal	a = b = c	$\alpha = \beta = \gamma \neq 90°$ $< 20°$	S	Al_2O_3, Calcite, $NaNO_3$
VII. Hexagonal	a = b ≠ c	$\alpha = \beta = 90°$ $\gamma = 120°$	S	Graphites, Ice, CuS, ZnO

* S stands for simple cubic (primitive unit cell)

$$= \sqrt{\frac{8RT}{\pi M}} = \sqrt{\frac{8KT}{\pi m}}$$

$$C = \sqrt{\frac{C_1^2 + C_2^2 + \ldots + C_n^2}{n}}$$

$$\alpha = \sqrt{\frac{2RT}{M}} = \sqrt{\frac{2KT}{m}}$$

$$\text{Translatory Kinetic energy/mole} = \frac{3}{2}RT = \frac{1}{2}MC^2$$

$$\lambda \doteq \eta\sqrt{\frac{3}{P\rho}} \simeq \frac{m}{\pi d^2 \rho} = \frac{m}{\sqrt{2}\pi d^2 \rho}$$

Gas Analysis

1. One mole of all gases occupy 22.4 litres at S.T.P.
2. Equal no. of moles occupy equal volume at the same temperature and pressure.
3. Volume of solid or liquid is considered negligible in comparison to the volume of a gas.

4. For the determination of formula of a hydrocarbon :

$$C_xH_y + \left(x + \frac{1}{4}y\right)O_2 = xCO_2 + \frac{1}{2}yH_2O$$

5. For one cc of hydrocarbon :
   - volume of CO_2 produced = x cc
   - volume of O_2 required = $x + \frac{1}{4}y$ cc
   - Contraction in volume = $1 + \frac{1}{4}$y cc

Absorbant	**Gas Absorbed**
Conc. H_2SO_4	moisture, NH_3
NaOH or KOH soln.	CO_2, SO_2, Halogens, NO_2
Ammoniacal cuprous chloride	CO, C_2H_2
Alkaline Pyragallol	O_2
Heated Palladium	H_2
Heated Magnesium	N_2
$FeSO_4$ Soln.	NO
Turpentine	O_3

Differences Between Ideal and Real Gases

Ideal gases	Real gases
I. No intermolecular forces of attraction.	I. Forces of attraction between molecules are significant.
II. Obey all gas laws.	II. Do not obey gas laws at high P and low T.
III. Obey ideal gas equation at all P,T.	III. Obey ideal gas equation at low P, high T and Vander Waals's equation at high P, low T.
IV. Size of molecules negligible as compared to distance between them.	IV. Finite size of molecules.
V. Whole volume of container is available for molecular movement.	V. Actual volume available for free movement of molecules is less than the volume of the container.
VI. Do not exist in nature.	VI. All gases are real.
VII. Volume becomes zero at absolute zero.	VII. Volume is not zero at absolute zero.

Close-packed Crystal Structure

The crystals have the atoms arranged such that minimum volume is left unoccupied. The two types of close-packed structures are:

(*a*) Hexagonal close packing (hcp)

(*b*) Cubic close packing or face-centred cubic close packing (fcc).

I. Hexagonal close packing (hcp)

(i) It has ABAB .... type of arrangement.

(ii) Spheres of third layer are placed exactly above the first and of fourth layer above the second.

(iii) Its coordination number is 12.

(iv) Its packing fraction is 74%.

II. Face centred cubic close packing (fcc)

(i) It has ABABC ... type of arrangement.

(ii) Spheres of third layer are placed on the interstices of the first two layers.

(iii) Its coordination number is 12.

(iv) Its packing fraction is 74%.

Differences Between Simple, Body Centred and Face-centred Cubic Systems

sc	*bcc*	*fcc*
I. One atom per unit cell.	I. Two atoms per unit cell.	I. Four atoms per unit cell.
II. Atomic radius, $r = a/2$ where a = cube edge.	II. Atomic radius, $r = \frac{\sqrt{3}}{4}a$	II. Atomic radius, $r = 2\sqrt{2}\ a.$
III. Packing fraction is 52%	III. Packing fraction is 68%	III. Packing frac -tion is 74%
IV. Coordination number is 6.	IV. Coordination number is 8.	IV. Coordination Number is 12.

Do You Know?

1. All gases are composed entirely of non-metallic elements and have low molecular mass.
2. Pressure, volume and temperature of a gas are related quantities. Any change in one affects the other two.
3. Pressure exerted by a gas on the walls of its container is defined as the force exerted

by the gas per unit area of the walls of the container.

4. Temperature of a gas indicates the kinetic energy of the gas molecules. Higher the temperature, higher is the kinetic energy and vice-versa.
5. Gas laws are independent of the nature of the gas.
6. Real gases do not obey gas laws at very low temperatures and very high pressures as they tend to lightly under such conditions.
7. Boyle's law asserts the compressibility and expansion of gases. At constant T, if pressure is increased, then volume of the gas will decrease i.e., it gets compressed. However, if pressure is released, the volume of the gas will increase, i.e., it expands.
8. According to Charles' law the absolute zero temp. (0K) is the lowest attainable temperature at which volume of an ideal gas would reduce to zero, i.e., it gets liquified at a constant pressure.
9. According to Gay-Lussac's law, the gas exerts no pressure at absolute zero, if volume is constant. This means that the

gas molecules freeze, i.e., molecular movement ceases at 0K.

10. 1 mole of any gas occupies a volume of 22.4 litres at S.T.P.
11. The universal molar gas constant R is independent of the amount or nature of the gas.
12. Air is a mixture of gases and contains about 21 parts oxygen and 79 parts nitrogen by volume, with traces of water.
13. All gases have same average kinetic energy at the same temperature.
14. All gas laws can be derived from the results of kinetic theory of gases.
15. Real gases tend towards ideality as temperature is increased and pressure is reduced.

6

CHEMICAL KINETICS

I. If $aA + bB \rightarrow cC$

Then instanteneous rate of reaction

$$= -\frac{1}{a}\frac{d[A]}{dt} = -\frac{1}{b}\frac{d[B]}{dt} = \frac{1}{c}\frac{d[C]}{dt}$$

II. If rate of reaction = $k\,[A]^l\,[B]^m\,[C]^n$

Then overall order of reaction = $l + m + n$

III. Units of rate constant (k) for

Zero order reaction = mol L^{-1} s^{-1}

First order reaction = s^{-1}

Second order reaction L mol^{-1} s^{-1}

Third order reaction = L^2 mol^{-2} s^{-1}

n^{th} order reaction = $(mol\ L^{-1})^{1-n}\ s^{-1}$

IV. $$K_1 = \frac{2.303}{t}\log\frac{[A_0]}{[A]} = \frac{2.303}{t}\log\frac{a}{a-x}$$

V. $$t_{1/2} = \frac{0.693}{k_1}$$

VI. $k = Ae^{-Ea/Rt}$ (Arrhenius equation)

VII. Rate of reaction is always a positive quantity.

VIII. Rate law is the experimentally determined expression that represents the relation between reaction rate and concentration of the reactants.

IX. Instantaneous rate of reaction at any time can be determined by the slope of the tangent at a point on concentration vs. time curve corresponding to that time.

X. If time interval is very small, then average rate of reaction equals instantaneous rate of reaction.

XI. Units of rate constant for gaseous reactions are:
For first order reaction = s^{-1}.
For second oreder reaction = $atm^{-1}\ s^{-1}$.
For third order reaction = $atm^{-2}\ s^{-1}$.

XII. Unit of rate constant for n^{th} order reaction are,

$$K = (mol\ L^{-1})^{1-n}\ s^{-1}$$

XIII. Collision frequency is the number of collisions that take place per second per unit volume of the reaction mixture.

XIV. Boltzamann factor = $e^{-Ea/RT}$.

XV. The catalyst does not change the heat or energy of reaction.

XVI. Einstein's law of photochemical equivalence states that absorption of one photon of light activates 1 atom. or 1 molecule.

XVII. Smaller the particle size, greater is the rate of reaction.

XVIII. First order reaction has a constant half-life.

XIX. The rate constant is a measure of the number of activated molecules per unit volume per second.

Law of Mass Action

1. For a reversible reaction : $A + B \rightleftharpoons C + D$,

$$K_C = \frac{[C][D]}{[A][B]}$$

2. For a general reversible reaction : $aA + bB \rightleftharpoons cC + dD$

$$K_C = \frac{[C]^c\,[D]^d}{[A]^a\,[B]^b} \text{ and } K_P = \frac{P_C^c \times P_D^d}{P_A^a \times P_B^b}$$

3. $K_P = K_C(RT)^{\Delta n}$
where, Δn = Change in no. of moles = No. of moles of product – No. of moles of reactants

If $\Delta n > 0$, $K_P > K_C$. $[PCl_5 \rightleftharpoons PCl_3 + Cl_2]$

If $\Delta n < 0$, $K_P < K_C$. $[N_2 + 3H_2 \rightleftharpoons 2NH_3]$

If $\Delta n = 0$, $K_P = K_C$. $[2HI \rightleftharpoons H_2 + I_2]$

where, K_C = concentration equilibrium constant

and K_P = Partial pressure equilibrium const.

- Rate of decay $\left(\frac{dN}{dt}\right) = -\lambda N$; where λ is called decay constant or disintegration constant. – ve sign indicates that N decrease with lapse of time.
- $T_A = \frac{1}{\lambda}$; where T_A is mean life or average life.
- $T_{1/2} = 0.693 \times T_A = 69.3\%$ of T_A; where $T_{1/2}$ is half life
- $T_{1/2} = \frac{0.693}{\lambda} = \frac{\log_e 2}{\lambda}$
- $\frac{N}{N_o} = \left(\frac{1}{2}\right)^n$

where, N_o = Original number or amount of nuclei,

N = Number of undisintegrated nuclei after lapse of time t

N = Number of half lives passed and $n = t/T_{1/2}$.

- % Radioactivity = $\left(\frac{1}{2}\right)^n \times 100$

- **Remember** : Full life > Average life > Half life

- **Unit of Radioactivity**

(a) Curie : It is the quantity of a radioactive substance which gives 3.7×10^{10} disintegrations /sec.

1 Curie = 3.7×10^{10} disintegrations/sec

1 Milli Curie (10^{-3} Curie) = 3.7×10^{7} disintegrations/sec

1 Micro Curie (10^{-6} Curie) = 3.7×10^{4} disintegrations/sec.

(b) Rutherford : It is the quantity of a radioactive material which gives 10^{6} disintegrations/sec.

1 Rutherford = 10^6 disintegrations/sec

1 Milli Rutherford = 10^3 disintegrations/sec

1 Micro Rutherford = 1 (one) disintegrations/sec

(c) **Becquerel** : It is the quantity of a radioactive element which gives 1 (one) disintegration per second.

- **Photo-electric effect** : The phenomenon of emission of electrons from the surface of metals when light radiations of suitable frequency fall on them, is called photo-electric effect.

Laws of Photo-electric Effect

(*i*) For a given metal, there exists a certain minimum frequency of the incident light radiations (called as threshold frequency) below which no emission of photo-electrons takes place.

(*ii*) The number of photo-electrons emitted per second (i.e., photo-electric current) is directly proportional to the intensity of the incident light and is independent of its frequency.

(*iii*) The emission of photo-electrons is an instantaneous process, i.e., it starts immediately as the light falls on the surface of the metal.

Factors Influencing Rate of Chemical Reactions

Following factors are influencing the rate of chemical reactions:

(*a*) *Concentration of the reactants:* Reaction rate increases with the increase of concentration of the reactatns.

(*b*) *Nature of reactants*

(*c*) *Temperature:* Reaction rate increase with the rise of temperature of the reactants.

(*d*) *Surface area:* The larger the surface area, the faster is the reaction rate.

(*e*) *Presence of a catalyst*: Catalyst increases the rate of reaction.

(*f*) *Exposure to radiation:* Generally, exposure to radiation increases the rate of reaction.

Concentration of the Reactants

Rate of chemical reaction increases with the increase of concentration of the reactants.

According to law of mass action, at the given temperture the rate of a chemical reaction is directly proporational to the product of the molar concentration of the reactants.

Consider the following general equation,

$$aA + bB \rightarrow \text{products}$$

According to the law of mass action (theoretical concept),

Rate of reaction = k $[A]^a$ $[B]^b$.

But the rate of reaction depends upon the experimentally determined value '*l*' concentration term of A and '*m*' concentration term of B.

But experimentally, the **rate of reaction** is given by the rate law as:

Rate of reaction = k $[A]^l$ $[B]^m$

Where, k = rate-constant

l and m = order of reaction with respect of A and B respectively.

Differences Between Rate of Reaction and Rate-constant

Rate of reaction	Rate-constant
I. Rate of reaction is the rate of disappearance of a reactant or rate of appearance of a product.	I. Rate constant is a proportionality constant in the rate law and is equal to rate of reaction when the concentration of reactant is unity.
II. It is dependent upon the concentration of the reactants.	II. It is independent of the initial concentration of the reactants.
III. Its unit is mol L^{-1} s^{-1}.	III. The unit of rate constant depends upon the order of the reaction.

Order of Reaction

Order of reaction is defined as the sum of powers of the concentration terms in the experimentally determined rate equation.

The order of above reaction = $l + m$

Depending upon whether $l + m$ is 0, 1, 2 or 3 the reaction is said to be zero order, Ist order, 2nd order or 3rd order.

I. **Zero order reaction:** When the reaction-rate is independent of the concentration of the reacting species, the reaction is known to be zero order reaction.

Here, $2\,HI\,(g) \rightarrow H_2\,(g) + I_2\,(g)$

Unit of rate constant (k) for zero order reaction:

$$\text{Rate} = k\,[A]^\circ = k$$

So, $k = \text{mol L}^{-1}\,\text{s}^{-1}$

II. **First order reaction:** When the reaction-rate is dependent upon the one concentration term of the reacting species, the reaction is known to be a first order reaction. Now.

$$A \rightarrow \text{Product}$$

$$\text{Rate} = k\,[A]$$

Here, $2N_2O_5\,(g) \rightarrow 4NO_2\,(g) + O_2\,(g)$

$$\text{Rate} = k\,[N_2O_5]$$

Unit of rate constant (k) for first order reaction:

$$\text{Rate} = k\,[A]$$

$$\text{mol L}^{-1}\,\text{s}^{-1} = k\,(\text{mol L}^{-1})$$

$$k = \text{s}^{-1}.$$

III. Second order reaction: When the reaction rate is dependent upon the two concentration terms of the reacting species, the reaction is said to be a second order reaction. Now,

$$A + A \rightarrow \text{product}$$

$$\text{Rate} = k\,[A]^2$$

Here, $2NO_2\,(g) + F_2\,(g) \rightarrow 2NO_2F\,(g)$.

$$\text{Rate of reaction} = k\,[NO_2]\,[F_2]$$

The order of raction = 1 + 1 = 2.

Unit of rate constant (k) for second order reaction:

$$\text{Rate} = k\,[A]^2$$

$$\text{mol L}^{-1}\,\text{s}^{-1} = k\,[\text{mol L}^{-1}]^2$$

$$k = \text{L mol}^{-1}\,\text{s}^{-1}$$

IV. Third order reaction: When the reaction rate is dependent upon the three concentration terms of the reacting species, the reaction is said to be a third order reaction. Now,

$$A + A + A \rightarrow \text{product}$$

$$\text{Rate} = k\,[A]^3$$

Here, $2NO\,(g) + O_2\,(g) \rightarrow 2NO_2\,(g)$

$$\text{Rate of reaction} = k\,[NO]^2\,[O_2]^1.$$

The order of reaction = 2 + 1 = 3.

Unit of rate constant (k) for third order reaction :

$$\text{Rate} = k\,[A]^3$$

$$\text{mol L}^{-1}\,\text{s}^{-1} = k\,[\text{mol L}^{-1}]^3$$

$$k = \text{L}^2\,\text{mol}^{-2}\,\text{s}^{-1}.$$

V. **Fractional order of reaction:** As the rate of reaction is dependant upon the experimentally determined concentration terms of reacting species and not on the stoichiometric coefficients in the balanced equation, so order of reaction can be fractional also.

e.g., $CH_3CHO(g) \xrightarrow{723\,K} CH_4(g) + CO(g)$

$$\text{Rate} = k\,[CH_3CHO]^{3/2}$$

7

THEORY OF DILUTE SOLUTIONS

I. Mole percent of a component

$$= \frac{\text{Number of moles of that component}}{\text{Total number of moles in solution}} \times 100$$

II. Molarity of the solution

$$= \frac{\text{Number of moles of the solute}}{\text{Volume of the solution in litres}}$$

III. Molality of the solution

$$= \frac{\text{Number of moles of the solute}}{\text{Mass of the solvent in kg}}$$

IV. Formality of the solution

$$= \frac{\text{Number of formula masses of the solute}}{\text{Volume of the solution in litres}}$$

V. Normality of the solution

$$= \frac{\text{Number of gram equivalents of the solute}}{\text{Volume of the solution in litres}}$$

VI. Normality = Molarity × $\frac{\text{Molar mass}}{\text{Equivalent mass}}$

VII. Normality (for acids) = Molarity × Basicity

VIII. Normality (for bases) = Molarity × Acidity

Types of Solutions

S.No.	Physical state of Solute	Solvent	Solution	Examples
I	Solid	Solid	Solid	Alloys geme etc
II	Liquid	Solid	Solid	Mercury in zinc or gold (amalgams) etc.
III.	Gas	Solid	Solid	Hydrogen in palladium etc.
IV.	Solid	Liquid	Liquid	Sugar in water, salt in wate etc.
V.	Liquid	Liquid	Liquid	Alcohol in water, benzene in toluene etc.
VI.	Gas	Liquid	Liquid	Oxygen in water, CO_2 in water etc.
VII.	Solid	Gas	Gas	Dust particles in air, iodine in air etc.
VIII.	Liquid	Gas	Gas	Water in air (like humidity) etc.
IX.	Gas	Gas	Gas	Mixture of various gases, air etc.

Solution, Solute and Solvent

A solution is a homogeneous mixture of two or more substances. A solution of two substances is called a binary solution. Its two components are called as solute and solvent.

A substance which is dissolved into another substance is called as **solute** and another substance which dissolves the solute into it is called as solvent. In sugar solution sugar is solute and water is **solvent.**

Solubility

The maximum amout of any solute that can be dissolved in 100 g of solvent at a particular temperature is called the solubility of that solute at particulare temperature. Which depends upon

(*i*) nature of solute

(*ii*) nature of solvent

(*iii*) temperature of the solution

(*iv*) pressure (for gases)

1. Normality (N) = $\frac{w}{W} \times \frac{1000}{E}$; where, V in ml.

2. $N = \dfrac{\text{gm. litre}^{-1}}{\text{Eq. wt}}$

$= \dfrac{\text{wt. of solute in gm./litre soln.}}{\text{Eq. wt. of solute}}$

3. Molarity (M) $= \dfrac{\text{gram. litre}^{-1}}{\text{Molecular wt.}}$

4. $M = \dfrac{w}{V} \times \dfrac{1000}{\text{Mol. wt.}}$; where, V in ml.

5. $V_1S_1 + V_2S_2 + V_3S_3 + ... = V_m \times S_m$.
For mixture containing all acids or alkalies.

6. $\{V_1S_1$ (acid one) $+ V_2S_2$ (acid two) $+ ....\} -$ $V'S'$(alkali one) $+ V''S''$(alkali two) $+ ....$
$= V_mS_m$; For mixture containing acids and alkalies.

7. Factor

$= \dfrac{\text{wt. of the solute taken}}{\text{wt. of the solute required for desired strength}}$

8. Actual Strength = Factor × Approximate Strength.

N.B.

(*a*) For mono-basic acid (HCl) or mono-acidic base (KOH); Molarity (M) = Normality (N);
[∵ Mol. wt. = Eq.wt.]

(*b*) For di-basic acid (H_2SO_4) or di-acidic base [$Ca(OH)_2$];
M = 2(N); [∵ Mol. wt. = 2 × Eq. wt.]

(*c*) For tri-basic acid (H_3SO_4) or tri-acidic base [$Al(OH)_3$]; M = 3(N);
[∵ Mol. wt. = 3 × Eq. wt.]

Osmotic Pressure (P)

1. $P = \frac{nRT}{V} = \frac{m}{M} \times \frac{RT}{V} = CRT$

where, n = no. of moles of solute
T = temp. in °K
V = volume in litres containing *x* gm mole of solute,
C = molar concentration of a solution
R = solution constant and is analogous to gas constant.

2. PV = RT; where V = vol. in litre containing 1 gm mole

3. $\frac{P_1V_1}{n_1T_1} = \frac{P_2V_2}{n_2T_2}$ (in case of two solutions)

4. For two isotonic solutions at the same temp., $\frac{V_1}{n_1} = \frac{V_2}{n_2}$

5. For two isotonic solutions, if $V_1 = V_2$ then $n_1 = n_2$ at the same temp.

6. In case of dissociation : $\frac{M_c}{M_o} = \frac{P_o}{P_c} = 1 + (n-1)\alpha$ where, α is degree of dissociation.

7. In case of association of molecules, $\frac{M_c}{M_o} = \frac{P_o}{P_c} = \left(\frac{n-1}{n}\right)\alpha$

where, o stands for observed or abnormal value and c stands for theoretical or calculated or normal value.

- In case of dissociation : Actual concentration $= \{1 + (n-1)\alpha\}$ C and

$$P_c = \{1 + (n-1)\alpha\}\,CRT$$

Lowering of Vapour Pressure (ΔP)

1. Lowering of vapour pressure (ΔP) = $P - P_s$ where, P = V.P. of pure solvent and P_s = V.P. of solution.

2. $$\frac{\Delta P}{P} = \frac{P - P_s}{P} = \frac{n}{n + N};$$

   $$\left[\frac{n}{n + N} = \text{mole fraction of the solute}\right]$$

3. $$\frac{\Delta P}{P} = \frac{P - P_s}{P} = \frac{m}{w} \times \frac{W}{M}$$

   where, M = mol. wt. of solute
   W = mol. wt. of solvent
   m = wt. of solute in gm. and
   w = wt. of solvent in gm.

   $$\frac{\Delta P}{P} = \frac{P - P_s}{P} = \text{Relative lowering of V.P.}$$

Elevation in Boiling Point (B.P.) (ΔT_b)

1. ΔT_b = B.P. of solution – B.P. of pure solvent
2. $\Delta T_b \propto C_m$ (Raoult's law)
   where, C_m is the molal or molar concentration of solution.

3. $\Delta T_b = \dfrac{K_b \times 10^3 \times w}{W \times M}$

where w = mass of solute in gm.
W = mass of solvent in gm
M = mol. wt of solute and
K_b = molal elevation constant of a solute

4. $K_b = \dfrac{RT^2}{10^3 \times L} = \dfrac{0.002T^2}{L}$

where, L = Latent heat
T = Boiling point and
R = 2 cals deg^{-1} $mole^{-1}$

5. $M = \dfrac{K_b \times 10^3 \times w}{\Delta T_b \times W}$

Depression in Freezing Point (ΔT_f)

1. ΔT_f = F.P. of pure solvent – F.P. of solution

2. $\Delta T_f = \dfrac{K_f \times 10^3 \times w}{W \times M}$

3. $K_f = \dfrac{0.002T^2}{L}$

where, K_f is molal depression const. of a solvent and T is F.P.

4. Mol. wt. (M) = $\dfrac{K_f \times 10^3 \times w}{W \times \Delta T_f}$

5. For same solution $\dfrac{\Delta T_b}{\Delta T_f} = \dfrac{K_b}{K_f}$

8

THERMO CHEMISTRY

Some Basic Definitions

1. **Exothermic reaction:** The reaction in which heat is evolved, is called exothermic reaction.
2. **Endothermic reaction:** The reaction in which heat is absorbed, is called endothermic reaction.
3. **System:** It is that specified space of the universe which is under theromodynamic observation. It may be homogeneous or heterogeneous.
4. **Surroundings:** The remaining part of the universe that is outside the sytem, is called its surrounding. They are separated from each other by a real or imaginary boundary.
5. **State variables or state functions:** It is a measurable physical property of the system whose value depends only upon the state of

the sytem and not upon the path followed to attain this state. Some common state functions are: pressure (P), volume (V), temperature (T), internal energy (E), entropy (S), enthalpy (H) and free energy (G). Out of these function, **two state functions are sufficient to describe the state of any thermodynamic system.**

6. **Isothermal process:** A process during which the temperature of the system remains constant, is known as isothermal process.
7. **Adiabatic process:** A process during which the total quantity of heat of the sytem remains constant i.e., the heat is not allowed to enter or leave the system, is known as adiabatic process.
8. **Isochoric process:** A process during which the volume of the system remains constant, is known as isochoric process.
9. **Isobaric process:** A process during which the pressure of the system remains constant, is known as isobaric process.
10. **Irreversible process:** A process in which the direction of the process can not be reversed, is known as irreversible process.

11. **Reversible process:** A process in which direction of the process can be reversed by an infinitesimally small change in the state of the system, is known as reversible process.
12. **Cyclic process:** A process in which a system after undergoing a number of changes returns back to its original state, is known as cyclic process.

First Law of Thermodynamics

It states that, energy can neither be created nor destroyed but can be transformed from one form to another form. The total energy of an isolated system remains constant. Mathematically,

$$\Delta E = Q + W \quad ...(i)$$

Where ΔE = change in internal energy

Q = heat given to the system

W = work done on the system

But $W = -P.\Delta V$

So, $\Delta E = Q - P.\Delta V \quad ...(ii)$

$\Rightarrow Q = \Delta E + P.\Delta V$

At constant volume, $\Delta V = 0$. So, $Q = \Delta E$

Hence, at constant volume, internal energy change is equal to heat of the system.

Enthaply (H) and Enthalpy Change (ΔH)

It is defined as the sum of internal energy and pressure-volume work of the system. So,

$$H = E + PV \quad ...(i)$$

Like internal energy, enthalpy is also a state function. Its absolute value cannot be measured hence it is measured as change between two states A and B of the process.

$$\Delta H = H_B - H_A$$

From eq. (*i*) $\Delta H = \Delta E + P.\Delta V + V.\Delta P$

$$= Q + V\Delta P$$

At constant pressure, $\Delta P = 0$. Hence $\Delta H = Q$

Hence, **at constant pressure, enthalpy change is equal to heat of the sytem.**

For an exothermic reaction, $\Delta H = -ve$

For an endothermic reaction, $\Delta H = +ve$

Its SI unit is J mol^{-1} or kJ mol^{-1}.

Characteristics of Enthalpy

(*i*) Enthalpy is a state function.

(*ii*) Enthalpy change is independent of the path followed.

(*iii*) Enthalpy is an extensive property and its value depends upon the amount of substace.

Entropy (S)

Entropy is a measure of the degree of disorder of a system. Disorder of a system is measured by the energy absorbed by the system to attain that disorder. Gaseous state is the most disordered and therfore has the highest entropy.

For a reversible process at constant temperature, the change in entropy is,

$$\Delta s = \frac{Q}{T}$$

where Q = heat absorbed or evolved

T = constant temperature

If heat is absorbed in the process, $\Delta s = +$ ve

If heat is evolved in the process, $\Delta s = -$ve.

SI unit of entropy is JK^{-1}.

Second aw of Thermodynamics

It states that the change in entropy for every spontaneous process must always be positive and hence the entropy of universe always increases.

Characteristic of Entropy

(*i*) Entropy is a state function. Hence entropy change

$$\Delta S = S_{\text{final state}} - S_{\text{initial state}}$$

(*ii*) Entropy change is independent of the path followed.

(*iii*) Entropy is an extensive property and its value depends upon the amount of substance.

(*iv*) Entropy of the universe alwasy increases.

(*v*) Entropy change for a cyclic process is zero.

(*vi*) Entropy change for a reversible process is zero.

(*vii*) Entropy change for an irreversible spontaneous process is always greater than zero. i.e. $\Delta S > 0$.

Free Energy (G)

Gibb's free energy or simply free energy (G) is defined as the maximum available energy of the system which can be converted into useful work.

$$G = H - TS$$

where, H = enthalpy of the system

S = entropy of the system

T = absolute temperature of the system

At constant temperature, free energy change is given by,

$$\Delta G = \Delta H - T\Delta S$$

This is called is **Gibbs Helmholtz equation.**

Conditions for Spontaneous Processes (i.e. ΔG to be negative)

The conditions for spontaneous processes and the effect of temperature.

$$(\Delta G = \Delta H - T\,\Delta S)$$

ΔH	Sign of TΔS	Sign of ΔG	Temperature (T)	Nature of process
–ve (i.e. exothermic)	+ve	–ve		Spontaneous
–ve (i.e. exothermic)	–ve	–ve	Low (T ΔS < ΔH)	Spontaneous
		+ve	High (T ΔS > ΔH)	Non-Spontaneous
+ve (i.e. endothermic)	–ve	+ve		Non-Spontaneous
+ve (i.e. endothermic)	+ve	+ve	Low (T ΔS < ΔH)	Non-Spontaneous
		–ve	High (T ΔS > ΔH)	Spontaneous
Zero	+ve	–ve		Spontaneous
Zero	–ve	+ve		Non-Spontaneous

So, a process or reaction takes place in such a way that free energy is minimum.

Criterion for Feasibility of a Chemical Process

Only those chemical process are feasible or spontaneous in which free energy decreases i.e. ΔG is negative. There are three possibilities for ΔG:

(*i*) **ΔG is Zero.** In this case, the process does not proceed in any direction and is said to be in **equilibrium state.**

(*ii*) **ΔG is positive.** In this case, the reaction is non-spontaneous.

(*iii*) **ΔG is negative.** In this case, the reaction is spontaneous.

Characteristics of free energy

(*i*) Free energy is a state function. Hence free energy change

$\Delta G = G_{\text{final state}} - G_{\text{initial state}}$

(*ii*) Free energy change is independent of the path followed.

(*iii*) Free energy is an extensive property and its value depends upon the amount of substance.

(*iv*) $\Delta G_{\text{system}} = T\Delta S_{\text{universe}}$

change gives the maximum
ained from a process.
s decreases (or free energy
ys negative) during a
ocess.

of thermodynamics

at, at absolute zero, the entropy of a
ly crystalline substance may be taken as
o.

Standard entropy change (ΔS°)

The entropy of one mole of substance at 298 K and 1 atm pressure is called as standard entropy (S°)

Standard entropy change in a reaction

= (sum of standard entropies of products)

– (sum of standard entropies of reactants)

or $\Delta S^\circ = \Sigma S^\circ$ (products) – ΣS° (reactants)

Limitations of thermodynamics

(*i*) It deals with microscopic systems and does not tell anything about macroscopic system.

(*ii*) It deals with only initial and fin.. the sytem.

Some Importent Results

1. Intrinsic energy = – Heat of formation

   = Enthalpy change of formation

2. For exothermic reaction :

   $\Delta H = H_P - H_R = -Q$ i.e., $H_P < H_R$

3. For endothermic reaction :

   $\Delta H = H_P - H_R = +Q$ i.e., $H_P > H_R$

4. ΔH_V be heat of reaction at constant vol. and ΔH_P be heat of reaction at constant pressure, then : $\Delta H_P = \Delta H_V + \Delta nRT$

   where, Δn = change in no. of moles = no. of moles of gaseous products – no. of moles of gaseous reactants.

   If $\Delta n > 0$, then $\Delta H_P > \Delta H_V$

   If $\Delta n < 0$, then $\Delta H_P < \Delta H_V$

   If $\Delta n = 0$, then $\Delta H_V = \Delta H_P$

5. Kirchoff's equation : $\Delta H_P = \Delta C_P (T_2 - T_1)$

9

IONIC EQUILIBRIUM, P^H AND HYDROLYSIS

Physical Equilibrium

It is the equilibrium between the same chemical species in different phases. e.g.

(*i*) Equilibrium between a liquid and its vapour.

(*ii*) Equilibrium between a vapour and its saturated solution.

Homogeneous Equilibrium

A reversible reaction having all the species in same phase throughout, is called in homogeneous equilibrium.

Here, $\underset{\text{Hydrogen}}{H_2(g)} + \underset{\text{Iodine}}{I_2(g)} \rightleftharpoons \underset{\text{Hydrogen iodide}}{2HI(g)}$

Here all reactants and products are gases.

Heterogeneous Equilibrium

A reversible reaction having its species in two or more phases, is called in heterogeneous equilibrium.

Here, $CaCO_3 (s) \rightleftharpoons CaO (s) + CO_2 (g)$

Calcium carbonate — Calcium oxide — Carbon dioxide

In this reaction $CaCO_3$ and CaO are solids, but CO_2 is a gas.

Ionic Equilibrium

A reversible reaction between species and its ions is called in ionic equilibrium.

e.g. $H_2O (l) \rightleftharpoons H^+ (aq) + OH^- (aq)$

Relationship between K_c and K_p

K_c = Equilibrium constant expressed in mol/litre.

K_p = Equilibrium constant expressed in pressures.

$$K_p = K_c (RT)^{\Delta n}$$

where R = 0.08206 litre atm K^{-1} mol^{-1}.

T = absolute temperature.

Δn = number of moles of the gaseous products – number of moles of the gaseous reactants in the balanced equation.

Le-Chatelier Principle

It states that if some change is introduced in a system at equilibrium, the system proceeds in such a way that the effect of change is minimised.

Factors influencing equilibrium

(*i*) Concentration of reactant or a product

(*ii*) Reaction volume or applied pressure.

(*iii*) Temperature

Electrolytes and non-electrolytes

Electrolyte: A compound whose aqueous solution or melt conducts electricity, is known as an electrolyte.

Non-electrolyte: A compound whose aqueous solution or melt does not conduct electricity, is known as a non-electrolyte.

Strong electrolyte: A substance which dissociates completely into its ions in an aqueous solution and hence is a very good conductor of electricity, is known as strong electrolyte.

Weak electrolyte: A substance which dissociates to a small extent is an aqueous solution, is known as weak electrolyte.

Degree of dissociation or ionisation (α): Fraction of electrolyte that dissociates into its ions when it is dissolved in water, is known as its degree of dissociation or ionisation, α is 1 for strong electrolytes and less than one for weak electrolytes.

Ionisation Constant of a Weak Electrolyte

According to this law, for a weak electrolyte, the degree of ionisation is inversely proportional to the square root of its molar concentration. It can be proved as follows –

Here,

$$AB + H_2O \rightleftharpoons A^+ (aq) + B^- (aq)$$

$$\Rightarrow \quad AB (aq) \rightleftharpoons A^+ (aq) + B^- (aq)$$

Moles before dissociation: 1 0 0

Moles after dissociation: (1–α) α α

If C mol/litre is initial concentration of the electrolyte AB.

$$[AB] = C(1-\alpha) \text{ mol/litre}$$
$$[A^+] = [B^-] = C\alpha \text{ mol/litre}$$

According to, law of equilibrium,

$$K = \frac{[A^+][B^-]}{[AB]}$$

$$= \frac{C\alpha \times C\alpha}{C(1-\alpha)} = \frac{C\alpha^2}{1-\alpha}$$

where K is ionisation constant for weak electrolytes. For weak electrolytes, $\alpha << 1$, so α can be neglected as compared to 1 *i.e.* $1 - \alpha \approx 1$.

$$\therefore \quad K = C\alpha^2$$

$$\Rightarrow \quad \alpha = \sqrt{\frac{K}{C}}$$

Arrhenius Concept of Acids and Bases

Acid: An acid is a substance which contains hydrogen and which when dissolved into water gives hydrogen ions (H^+). The acid ionizes completely when dissolved in water, is called strong acid. An acid ionizes partially when dissolved in water, is called **weak acid.**

A **weak acid** solution contains unionized molecules in addition to ions. In general,

$$\underset{\text{Acid}}{HA\,(aq)} \rightleftharpoons H^+\,(aq) + A^-\,(aq)$$

Base: A base is a substance which contains hydroxyl group and which when dissolved into water gives hydroxyl ions (OH^-).

A base that ionizes completely into its ions when dissolved in water, is called a **strong base**, and a base ionizes partially when dissolved in water is called a **weak base**. In general,

$$\underset{\text{Base}}{BOH\,(aq)} \rightleftharpoons B^+\,(aq) + OH^-\,(aq)$$

Neutralisation is the process in which hydrogen ions and hydroxyl ions combine to form unionized molecules of water.

$$H^+\,(aq) + OH^-\,(aq) \rightleftharpoons H_2O\,(l)$$

Bronsted-Lowry Concept of Acids and Bases

An **acid** is defined as a substance which has the tendency to give a proton (H^+).

A **base** is defined as a substance which has a tendency to accept a proton (H^+).

Conjugate base: The remainder part of an acid after donating a proton, is called conjugate base.

$$\underset{\text{Acid}}{HCl\,(g)} + H_2O\,(l) \rightarrow H_3O^+\,(aq) + \underset{\text{Conjugate base}}{Cl^-(aq)}$$

Conjugate acid: Base when accepts a proton released by an acid is called as conjugate acid.

$$\underset{\text{Base}}{H_2O\,(l)} + H^+\,(\text{from an acid}) \rightarrow \underset{\text{Conjugate acid}}{H_3O^+\,(aq)}$$

Amphoterism: A particular can behave as an acid by donating a proton in one reaction and as a base in another by accepting a proton. Such a species is called as an amphoteric species. such as water.

$$\underset{\text{Acid}_1}{H_2O} + \underset{\text{Base}_2}{H_2O} \rightleftharpoons \underset{\text{Base}_1}{OH^-} + \underset{\text{Acid}_2}{H_3O^+}$$

Lewis Concept of Acids and Bases

Acid : An acid is a substance or an ion which is capable of accepting a pair of electrons and is called as Lewis acid.

Base : Base is a substance or an ion which is capable of donating an unshared pair of electrons and is called as Lewis base.

Types of Lewis Bases

(*i*) *Neutral molecules* having at least one lone pair of electrons. e.g. NH_3, R–NH_2, R_2–NH, R–OH, H–OH etc.

(*ii*) All negative ions like F^-, Cl^-, Br^-, OH^-, CN^- etc.

Types of Lewis Bases

(*i*) Electron deficient compounds or molecules having a central atom with incomplete octet : *e.g.* BF_3, $AlCl_3$ etc.

(*ii*) Cations e.g. Ag^+, Cu^{2+}, Fe^{3+}, etc.

(*iii*) Molecules with multiple bonds between two atoms of different electronegativities *e.g.* CO_2.

(*iv*) Molecules whose central atoms has empty d-orbitals, *e.g.* $SnCl_4$, SiF_4, PF_5, PCl_5 etc.

Some Important Results

1. $P^H = -\log [H^+]$
2. $[H^+] = 10^{-P^H}$
3. $P^{OH} = -\log [OH^-]$
4. $P^{Ka} = -\log K_a$
5. $p^{K_b} = -\log K_b$

6. $p^{K_w} = -\log K_w$
7. P^H of acidic solution < 7
8. P^H of neutral solution $= 7$
9. P^H of basic solution > 7
10. $P^H + P^{OH} = P^{K_w} = 14$
11. K_w is called ionic product of water $= [H^+]\,[OH^-] = 10^{-14}$ at 25°C
Thus in pure H_2O; $[H^+] = [OH^-] = 10^{-7}$ molar
12. $K_h = \frac{K_w}{K_a}$, for hydrolysis of anion of weak acid

$K_h = \frac{K_w}{K_b}$, for hydrolysis of cation of weak base.

$K_h = \frac{K_w}{K_a . K_b}$, for hydrolysis of weak acid-weak base salt.

where, K_h is hydrolysis constant, K_a is ionisation constant for an acid otherwise called acid dissociation constant and K_b is ionisation constant for a base otherwise called base dissociation constant.

13. **P^H of an acidic buffer solution and P^{OH} of an alkaline buffer solution.**

(a) Acidic Buffer : (Weak acid and its salt which is a strong electrolyte; e.g., CH_3COOH and CH_3COONa); in general, $H \rightleftharpoons HA^+ + A^-$

$P^H = P^{K_a} + \log \frac{[A^-]}{[HA]}$ (Henderson's equation)

(b) Basic Buffer : (Weak base and its salt which is a strong electrolyte; e.g., NH_4OH and NH_4Cl) in general,

$$BOH \rightleftharpoons B^+ + OH^-, P^{OH} = P^{K_b} + \log \frac{[B^-]}{[BOH]}$$

10

REDOX REACTIONS

Term	electron change	Oxidation number change
Oxidation	Loss of electrons	Increases
Oxidizing agent	Takes electrons	Decreases
Reduction	Gain of electrons	Decreases
Reducing agent	Gives electrons	Increases

A Classical Concept of Oxidation and Reduction

A process which involves addition of oxygen or any electronegative element or removal of hydrogen or any electropositive element is called oxidation.

A process which involves addition of hydrogen or any electropositive element or removela of oxygen or any electronegative element is called Reduction.

An Electronic Concept of Oxidation and Reduction

Oxidation is the process which involves loss of electrons.

$$Na \xrightarrow{\text{Oxidation}} Na^{+} + e^{-} \quad [0 \rightarrow +1]$$

Sodium atom — Sodium ion — Electron — Increase in +ve charge

$$Cl^{-} \xrightarrow{\text{Oxidation}} Cl + e^{-} \quad [-1 \rightarrow 0]$$

Chloride ion — Chlorine atom — Electron — Increase in +ve charge

Oxidation reaction can be known as it involves increase in positive charge or decrease in negative charge of the atom.

Reduction is the process which involves gain of electrons.

$$Fe^{3+} + 3e^{-} \rightarrow Fe \quad (+3 \rightarrow 0)$$

Ferric ion — Electrons — Iron — Decrease in +ve charge

$$O + 2e^{-} \longrightarrow O^{2-} \quad (0 \rightarrow -2)$$

Oxygen — Electrons — Oxide ion — Decrease in –ve charge

Reduction reaction can be known as it involves increase in negative charge or decrease in positive charge of the atom.

Oxidizing agent : Oxidizing agent oxidizes other substances on the expense of self reduction or it is the agent that gains electrons.

Reducing agent : Reducing agent reduces other substances on the expense of self oxidation or it is the agent that loses electrons.

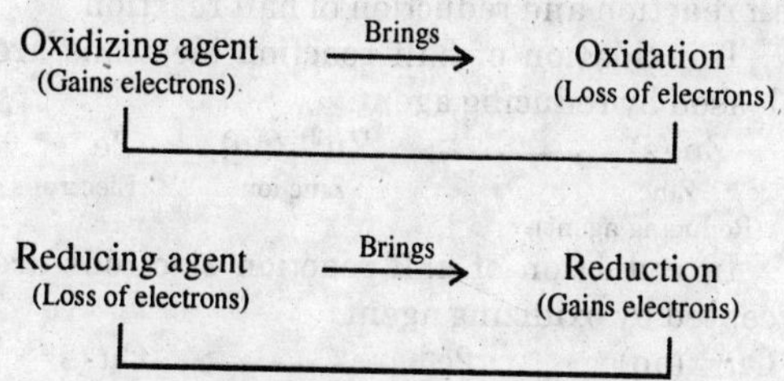

Redox reaction : It is the reaction in which oxidation and reduction occur simultaneously. A redox change involves the process in which a reducing agent is oxidized to liberate electrons, which are then taken up by an oxidizing agent to get itself reduced.

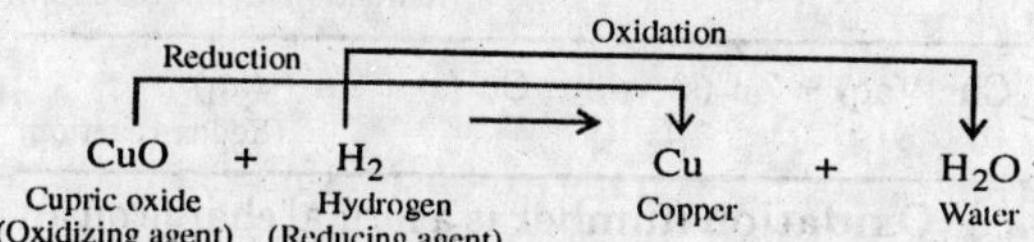

Oxidation Half Reaction and Reduction Half Reaction

Redox reaction is a combination of oxidation of half reaction and reduction of half reaction.

In oxidation of half reaction electrons are released by reducing agent.

$$\underset{\substack{\text{Zinc}\\\text{(Reducing agent)}}}{Zn\,(s)} \longrightarrow \underset{\text{Zinc ion}}{Zn^{2+}\,(aq)} + \underset{\text{Electrons}}{2e^-}$$

In reduction of half reaction electrons are accepted by oxidizing agent.

$$\underset{\substack{\text{Cupric ion}\\\text{(Oxidizing agent)}}}{Cu^{2+}\,(aq)} + \underset{\text{Electrons}}{2e^-} \longrightarrow \underset{\text{Cupper}}{Cu\,(s)}$$

Hence, Redox reaction is obtained by adding oxidation of half reaction and reduction of half reaction.

$$Zn(s) \rightarrow Zn^{2+}\,(aq) + 2e^-$$

(Oxidation of half reaction)

$$Cu^{2+}\,(aq) + 2e^- \rightarrow Cu\,(s)$$

(Reduction of half reaction)

$$Cu^{2+}\,(aq) + Zn\,(s) \rightarrow Cu\,(s) + Zn^{2+}\,(aq)$$

(Redox reaction)

Oxidation number is a formal charge on an atom in a compound or ion. Which is given positive given positive sign if electrons are lost and negative sign when electrons are gained.

Oxidation state of an element is the oxidation number per atom.

Rules to calculate oxidation number of an element

I. In elementary state, the atoms are assigned an oxidation number zero.

II. In compounds, oxidation number of

(a) Hydrogen is +1. In metallic hydrides oxidation number of hydrogen is –1.

(b) Oxygen is –2. In peroxides it is –1 and in F_2O it is +2.

(c) Alkali metals is +1 and alkaline earth metals is +2.

(d) Fluorine is always –1.

III. In neutral molecules, the sum of the oxidation numbers of all the atoms is zero.

IV. The oxidation number of simple ion is equal to the charge on the ion.

V. For charged species, the sum of the oxidation numbers of all the atoms is equal to the charge on the ion.

(a) Calculate the oxidation number of Cr in $K_2Cr_2O_7$.

$K_2Cr_2O_7$: $2 \times (+1) + 2(x) + 7(-2) = 0$

Hence, $x = +6$

Oxidation number of Cr in $K_2Cr_2O_7$ will be +6.

(b) Calculate the oxidation number of Mn in $KMnO_4$.

$KMnO_4 : 1 + x + 4(-2) = 0$

$x = +7$

Oxidation number of Mn in $KMnO_4$ will be +7.

Differences Between Oxidation Number and valency

Oxidation number	Valency
I. It is the formal charge on an atom in its combined state.	I. It is the number representing the combining capacity of an element with other elements.
II. It can be zero.	II. It is never zero except in case of noble gases.
III. It can be positive or negative.	III. It does not carry any charge, hence no plus or minus sign.
IV. It may be whole number or fractional.	IV. It is always a whole number.

Methods used to balance oxidation-reduction equations are :

I. Oxidation number method.

II. Ion electron method.

I. Oxidation number method :

(a) Identify the atoms which show change in oxidation number.

(b) Balance the oxidation number by multiplying it with a coefficient which makes the increase in oxidation number for the oxidized substances equal to the total decrease in oxidation number of reduced substances.

(c) Balance the charge by H^+.

(d) Balance the oxygen atoms by adding water and H^+ ions in acidic medium or by adding H_2O and OH^- ions in basic medium.

(e) Balance the elements which are not oxidized or reduced.

Now Balance the following redox equation :

$CuO + NH_3 \rightarrow Cu + N_2 + H_2O$

Step I: $\underset{+2}{CuO} + \underset{-3}{NH_3} \rightarrow \underset{0}{Cu} + \underset{0}{N_2} + H_2O$

Step II: Multiply CuO by 3 and NH_3 by 2.

$3CuO + 2NH_3 \rightarrow Cu + N_2 + H_2O$

Step III : Balance oxygen by multiplying H_2O by 3 and put 3 before Cu.

$3CuO + 2NH_3 \rightarrow 3Cu + N_2 + 3H_2O$

II. Ion electron method

(a) Select the oxidant and reductant.

(b) Write the oxidation half reaction and balance the atoms of other atoms (except H and O), O atoms (by adding water), H atoms (by H^+) and also charge (by electrons).

(c) Proceed in similar manner with reduction half reaction.

(d) Multiply these reactions with suitable co-efficient so as to balance electrons appearing on either side of these reactions.

(e) Add these two half reactions.

(f) If the reaction is taking place in basic solution, add enough OH^- on both sides of the half reaction to get rid of H^+. Combine H^+ and OH^- to give H_2O and cancel any duplication.

Now, Balance fhe following redox equation.

$Cr_2O_7^{-2} + C_2O_4^{2-} \rightarrow Cr^{3+} + CO_2$ in acidic medium

Step I : Oxidation half reaction

$C_2O_4^{2-} \rightarrow 2CO_2 + 2e^-$

Step II : Reduction half reaction

$Cr_2O_7^{2-} + 14H^+ + 6e^- \rightarrow 2\,Cr^{3+} + 7H_2O$

Step III : Add the oxidation half reaction and reduction half reaction

$$\begin{array}{rcl} Cr_2O_7^{2-} + 14\,H^+ + 6e^- & \rightarrow & 2\,Cr^{3+} + 7H_2O \\ C_2O_4^{2-} & \rightarrow & 2\,CO_2 + 2e^-] \times 3 \\ \hline Cr_2O_7^{2-} + 14\,H^+ + 3C_2O_4^{2-} & \rightarrow & 2Cr^{3+} + 7H_2O + 6CO_2 \end{array}$$

Oxidation Number and Nomenclature

Stock system of nomenclature : In this system, oxidation number of the element is indicated by a Roman numeral written in the bracket immediately after the name of the element.

Compound	Oxidation Number	Name of the compound According to stock system	Name of the compound Acording to old method
$FeSO_4$	Fe–2	Iron (II) sulphate	Ferrous Sulphate
$Fe_2(SO_4)_3$	Fe–3	Iron (III) sulphate	Ferric Sulphate
Cu_2O	Cu–1	Copper (I) oxide	Cuprous oxide
Cr_2O_3	Cr–3	Chromium (III) oxide	Chromium trioxide
$K_2Cr_2O_7$	Cr–6	Potassium dichromate (VI)	Potassium dichromate

Mn_2O_7	Mn–7	Manganese (VII) oxide	Manganese heptoxide
$SnCl_2$	Sn–2	Tin (II) chloride	stannous chloride
$SnCl_4$	Sn–4	Tin (IV) chloride	stannic chloride

Types of cells : Their differences

Electrochemical cell (or galvanic cell or valtaic cell)	Electrolytic cell
I. It ia a device used to convert chemical energy into electrical energy	I. It ia device used to convert electrical energy into chemical energy.
II. Oxidation occurs at anode (negative terminal) and reduction occurs at cathode (positive terminal).	II. Oxidation occurs at cathode (negtive) and reduction occurs at anode (positive).
III. The two electrodes are placed in different containers called half cells, connected through salt bridge.	III. Both the electrodes are placed in the same container.
IV. Work is obtained from the cell.	IV. Work is done on the cell.
V. Free energy decreases as the cell reaction proceeds.	V. Free energy increases as the cell reaction proceeds.
VI. Cell reaction is spontaneous.	VI. Cell reaction is non-spontaneous.

11

ELECTRO-CHEMISTRY

1. Equivalent conductance, $\Lambda = \dfrac{\sigma \times 1000}{C}$
2. Molar conductance, $\Lambda_m = \dfrac{\sigma \times 1000}{M}$
3. According to Kohlrausch's law,

$$\Lambda_m^\infty = n_+ \Lambda_{m+}^\infty + n_- \Lambda_{m-}^\infty$$

4. $E\left(M^{+n}/M\right) = E^\circ\left(M^{n+}/M\right) - \dfrac{2.303\,RT}{nF}$

$$\log \frac{[M(s)]}{\left[M^{n+}(aq)\right]} \text{ (Nerns't equation)}$$

5. $E^\circ_{cell} = 2.303 \dfrac{RT}{nF} \log K_c$
6. $\Delta G^\circ = -nF\,E^\circ_{cell} = -2.303\ RT \log K_c$
7. $m = WIt$ (Faraday's law of electrolysis)

Conductors and insulators

The substances which permit the passage of electric current through them, are called as conductors. But the substances which do not permit the passage of electric current through them, are called as non-conductors or insulators. Metals are good conductors of electricity.

Electrolytes and non-electrolytes

The substances which permit the passage of electric current through their molten state or in solution and are decomposed by it, are known as **electrolytes**. This phenomenon of passage of electric current through an electrolytic solution is known as **electrolytic conduction**. While, the substances which do not permit the passage of electric current through their molten state or in solution and are not decomposed by it, are known as **non-electrolytes**.

I. **Specific resistance or resistivity:** Specific resistance or resistivity of the material of a conductor is defined as the resistance of unit length and unit area of cross-section of that conductor.

Now specific resistance or resistivity is given as,

$$\rho = R\frac{A}{l}$$

Whare R = Resistance of the conductor
A = Area of cross-section of the conductor
l = Length of the conductor.

Its SI unit is ohm metre (Ωm).

II. Conductance: The reciprocal of the resistance of a conductor is called its conductance. It is denoted by G.

So,

$$G = \frac{1}{R}$$

Its SI unit is mho or ohm^{-1} (Ω^{-1}).

III. Specific conductance or conductivity: The reciprocal of the resistivity of the material of a conductor is called its conductivity. It is denoted by σ. Hence,

$$\sigma = \frac{1}{\rho}$$

Its SI unit is ohm^{-1} metre^{-1} ($\Omega^{-1}\, m^{-1}$).

IV. Equivalent conductance: Equivalent conductance is defined as the conductance of a volume of solution containing one gram equivalent of an electrolyte in solution. It is denoted by symbol Λ. Its unit is ohm^{-1} cm^2 $(g\text{-}equiv)^{-1}$.

Now, $$\Lambda = \frac{\sigma \times 1000}{C}$$

Where σ = specific conductance.
C = equivalents of solute per litre of solution.

V. Molar conductance is defined as the conductance of a volume of solution containing one gram mole of an electrolyte in solution. It is denoted by Λ_m. Its unit is ohm^{-1} cm^2 mol^{-1}.

Now, $$\Lambda_m = \frac{\sigma \times 1000}{M}$$

Where σ = specific conductance.
M = concentration of the solution in moles per litre.

VI. Electrolytic cell: It is a device used to convert electrical energy into chemical energy. Reduction takes place at cathode and oxidation takes placed at anode. When electricity is passed through electrolyte, it dissociates into its ions, cations move towards the cathode and anions move towards the anode. When electricity is passed through molten sodium chloride, then,

Electrochemical reaction, $NaCl \rightleftharpoons Na^+ + Cl^-$

At cathode : $Na^+ + e^- \rightarrow Na$ (reduction)

At anode : $Cl^- - e^- \rightarrow Cl$ (oxidation)

$Cl + Cl \rightarrow Cl_2$

VII. Faraday's laws of electrolysis:

First law of electrolysis: During electrolysis, the mass of the substance produced or consumed at an electrode is directly proportional to the quantity of electricity passed through the electrolyte.

Hence, $m = WIt$

Second law of electrolysis: When the same quantity of electricity is passed through different electrolytes connected in series, the weight of different substances produced at the electrodes are proportional to their equivalent weight.

VIII. Commercial cell or batteries: A battery is a series combination of two or more electrochemical cells used as a source of electrical energy. Following are the types of commercial cells.

Commercial cell

- Primary cells (They cannot be recharged) e.g. dry cell, mercury cell.
- Secondary cells or storage cells or accumulators (They can be reacharged) e.g. lead storage cell, nickel cadmium cell, fuel cells.

IX. **Electrolytic corrosion:** Corrosion is the gradual deterioration of metal by atmospheric attack. It is an electrochemical reaction. Corrosion of iron is called as rusting. Rust is mainly a mixture of ferric hydroxide [$Fe(OH)_3$] and ferric oxide [Fe_2O_3.].

Factors Affecting Molar Conductance

(a) **Nature of electrolytes:** If greater the number of ions in the solution, then greater is the conductance. Strong electrolytes dissociate completely into its ions and give large number of ions, therefore have high conductance. Weak electrolytes ionize to small extent and give lesser number of ions, so have low conductance.

(b) Concentration of the solution: Molar conductance of an elctrolyte increases with decrease in concentration.

For strong electrolytes, variation of molar conductance with concentration is given as,

$$\Lambda_m = \Lambda_m^{\infty} - b\sqrt{C}$$

Where Λ_m^{∞} = molar conductance at infinite dilution

C = concentration

b = a constant

(c) Temperature: The conductance of electrolytes increases with increase in temperature.

Kohlrausch's Law of Independant Migration of Ions

According to this law *at infinite dilution, if dissociation for all electrolytes is complete and if all inter-ionic effects disappear each ion migrates independently, and contributes to the total equivalent conductance of an electrolyte a definite share which depends only on its own nature and not at all on that of the ion with which it is* *associated. So, Λ_m of any electrolyte is the sum of the molar conductances of the ions composing it. Molar conductances of cations and anions at infinite dilution are represented as*

Λ^{∞}_{m+} and Λ^{∞}_{m-} respectively. Now,

$$\Lambda^{\infty}_{m} = n_+ \Lambda^{\infty}_{m+} + n_- \Lambda^{\infty}_{m-}$$

where n_+ and n_- are the number of cations and anions per formula unit of electrolyte.

For NaCl, $\Lambda^{\infty}_{m}(\text{NaCl}) = \Lambda^{\infty}_{m+}\left(\text{Na}^+\right) + \Lambda^{\infty}_{m-}\left(\text{Cl}^-\right)$.

For $Al_2(SO_4)_3$, $\Lambda^{\infty}_{m}\left[Al_2(SO_4)_3\right] = 2\Lambda^{\infty}_{m+}\left(Al^{3+}\right)$

$$+ 3\Lambda^{\infty}_{m-}\left(SO_4^{2-}\right)$$

Differences Between Electrochemical Cell and Electrolytic Cell

Electrochemical Cell (or galvanic cell or valtaic cell)	Electrolytic cell
I. It is a device used to convert chemical energy into electrical energy	I. It is a device used to convert electrical energy into chemical energy.

II. Cell reaction is spontaneous.	II. Cell reaction is non-spontaneous.
III. Free energy decreases as the cell reaction proceeds.	III. Free energy increases as the cell reaction proceeds.
IV. Work is obtained from the cell.	IV. Work is done on the cell.
V. The two electrodes are placed in different containers called *half cells*, connected through *salt bridge*.	V. Both the electrodes are placed in the same container.
VI. Oxidation occurs at anode (negative terminal) and reduction occurs at cathode (positive terminal).	VI. Oxidation occurs at cathode (negative) and reduction occurs at anode (positive).

Electrochemical Cell

In electrochemical cell, electricity is produced by spontaneous chemical reaction. In redox-reaction simultaneous loss and gain of electrons takes place. The electrons lost by reducing agents are taken up by oxidising agent, which causes the transfer of electrons and produced electricity. Electrochemical cells are also called as voltaic cells or galvanic cells.

Voltaic cell consists of zinc rod dipped in 1.0M copper sulphate solution. Electrons pass from zinc atoms to copper ions in which they are in direct contact with each other.

Oxidation half reaction:

$$Zn\ (s) \rightarrow Zn^{2+}\ (aq) + 2e^-$$

Reduction half reaction:

$$Cu^{2+}(aq) + 2e^- \rightarrow Cu(s)$$

Overall reaction:

$$\underset{\text{Zinc}}{Zn(s)} + \underset{\text{Cupric ion}}{Cu^{2+}(aq)} \rightarrow \underset{\text{Zinc ion}}{Zn^{2+}(aq)} + \underset{\text{Copper}}{Cu(s)}$$

Hence, It is represented as Zn (s) | Zn^{2+} (1M) || Cu^{2+} (1M) | Cu (s).

Daniel cell: It consists of two half cells, anode half cell and cathode half cell. The anode half cell consists of zinc electrode (anode) immersed in zinc sulphate solution and cathode half cell consists of copper electrode (cathode) immersed in copper sulphate solution. These two electrodes are connected with a copper wire through ammeter. The two solutions are connected through salt bridge.

At anode:

$$Zn(s) \rightarrow Zn^{2+}\ (aq) + 2e^- \quad \text{(oxidation)}$$

At cathode:

$$Cu^{2+}\ (aq) + 2e^- \rightarrow Cu\ (s) \qquad \text{(reduction)}$$

Electrochemical reaction:

$$Zn(s) + Cu^{2+}(aq) \rightarrow Zn^{2+}(aq) + Cu(s).$$

As the reaction starts, the mass of zinc rod decreases and that of copper rod increases.

Salt bridge: Salt bridge contains an inert paste of a strong electrolyte (Na_2SO_4 or KCl) and agar gel. Which prevents the mixing of two electrolytes and allows the migration of ions from cathode half cell to anode half cell.

Electrode Potential and E.M.F. of a Cell

Electrode potential: Electrode potenital is defined as the tendency of a metal to lose or gain electrons when placed in contact with its ions. It is of two types :

(a) Oxidation potential: The tendency of an electrode to lose electrons is known as its oxidation potential.

$Zn \rightleftharpoons Zn^{2+}(aq) + 2e^-$

(b) Reduction potential: The tendency of an electrode to gain electrons is known as its reduction potential.

$Cu^{2+}(aq) + 2e^- \rightleftharpoons Cu$

Oxidation potential is equal and opposite of reduction potential.

The electrode potential depends upon

(*a*) Concentration of metal ions in solution.

(*b*) Nature of the metal and its ions.

(*c*) Temperature.

E.M.F. or Cell Potential

The difference between the electrode potentials of the two electrodes providing an electrochemical cell is called, as electromotive force (E.M.F.) or cell potenital of a cell. Which is represented as E_{cell} and expressed in volts.

$$E_{cell} = E_{cathode} - E_{anode}$$

Distinctions Between E.M.F. and Potential Difference

	E.M.F		Potential difference
I.	It is the potential difference between the two electrodes when no current is being drawn from the cell i.e., in open circuit.	I.	It is the difference of potentials of two electrodes when current is being drawn from the cell i.e., in closed circuit.
II.	It is measured with the help of a potentiometer.	II.	It is measured with the help of a voltmeter.

III.	It is the maximum voltage that can be drawn from the cell.	III.	It is less than the E.M.F. of the cell.
IV.	It is a measure of maximum useful work obtainable from the cell.	IV.	It gives work which is less than the maximum obtainable work.

Electrode Potential, Electrolysis and E.M.F.

(*i*) $M^{n+} + ne = M$

(*ii*) $E_{OX.} + E_{Red} =$ e.m.f. (of cell) in volt

(*iii*) $\Delta G = -nEF.$

where, ΔG = Change in free energy

n = no. of electrons used in the reaction

E = e.m.f. of cell and

F = Faraday's constant = 96,500 coulombs

(*iv*) $\log K_e = \dfrac{nEF}{2.303RT}$, At 25°C, T = 273 + 25

= 298K and

$\dfrac{2.303}{F} = 0.059$ so, at 25°C, $\log K_e = \dfrac{nE}{0.059}$

where, K_e = equilibrium constant
E = e.m.f. of cell and
n = no. of electrons used in the reaction.

(*v*) Nernst equation: For $M_1^{n+} + M_2 \rightleftharpoons M_1 + M_2^{n+}$

$E_{ox} = E^0_{ox} - \frac{2.303RT}{nF} \log K_e$; where, E_{ox} = Oxidation potential and E^0_{ox} = standard oxidation potential

(*vi*) $W = ZQ = Z.i.t$ (Ist law of electrolysis)

(*vii*) $Z = \frac{E}{F}$ $\therefore W = \frac{E}{F} i.t.$

(*viii*) $1F = 96,500$ coulomb $= N \times e = 6.023 \times 10^{23} \times 1.6 \times 10^{-19}$ coulombs

(*ix*) $\frac{W_A}{W_B} = \frac{E_A}{E_B}$ (2nd law of electrolysis)

(*x*) $\frac{Z_A}{Z_B} = \frac{E_A}{E_B}$; where, Z is called electro-chemical equivalent and E is called chemical equivalent of the substance.

(*xi*) Oxidation occurs at anode.

(*xii*) Reduction occurs at cathode.

(*xiii*) 1F ≡ 1 gram equivalent of a substance
= 11,200 ml H_2 at N.T.P. = 5,600 ml O_2 at N.T.P.

(*xiv*) 1 Ampere = 1 Coulomb × 1 sec.

(*xv*) Unit of Z is the unit of mass per unit charge.

12

SURFACE CHEMISTRY

1. **Adsorption.** It is the phenomenon of higher concentration of molecules of gases or liquid on the surface than in the bulk of the solid.
2. **Types of adsorption**
   (*i*) Physical adsorption
   (*ii*) Chemical adsorption
3. **Types of solutions**
   (*i*) True solutions
   (*ii*) Colloidal solutions
   (*iii*) Suspensions
4. **Types of colloids**
   (*i*) Lyophillic colloids
   (*ii*) Lyophobic colloids
5. **Coagulation.** It is the phenomenon of precipitiation of a colloidal solution by the addition of excess of an electrolyte.
6. **Emulsions.** These are the colloidal solutions of two immiscible liquids.

7. **Gel.** It is the colloidal solution of a liquid in a solid.
8. **Dialysis:** In this method, the colloidal solution is taken in a bag made of cellophane or parchment paper and is suspended in fresh water. The impurities diffuse out of the bag leaving behind pure colloidal solution in the bag.
9. **Ultrafiltration:** In this process impurities from colloidal solution are removed by passing it through ultrafilter papers.
10. **Sorption** is the phenomenon in which adsorption and absorption occur together.
11. Physical adsorption is also called as **physisorption** and chemical adsorption is also called as **chemisorption**.
12. The reverse process of adsorption is called as **desorption**.
13. Adsorption is accompanied with evolution of heat, whereas desorption takes place with absorption of heat.
14. A lyophillic sol in which the dispersion medium is water, is called as **hydrophillic sol** or **emulsoid**.

15. A lyphobic sol in which the dispersion medium is water, is called as **hydrophobic sol or suspensoid.**
16. Gold sol obtained by the reduction of $AuCl_3$ by $SnCl_2$ in purple in colour and is called as **purple of cassius.**
17. The process of breaking an emulsion to get two immiscible liquids is called as **demulisification.**
18. Heterogeneous catalysis is also called as **surface catalysis.**

Distinction Between Adsorption and Absorption

Adsorption	Absorption
I. The concentration of adsorbate is higher on the surface of adsorbent.	I. The concentration of gas or liquid particles is uniform throughout the substance.
II. The rate of adsorption decreases slowly as the process proceeds.	II. Absorption rate remains uniform . throughout.
III. It is a surface phenomenon.	III. It is a bulk phenomenon.

Types of Adsorption

The two types of adsorption are, (i) physical adsorption and (ii) chemical adsorption.

Distinction between physical and chemical adsoptions

Physical adsorption	Chemical adsortion
I. Adsorbate is held with adsorbent by Vander Waal's forces.	I. Adsorbate are held with adsorbent by strong chemical forces.
II. It forms multimolecular layers.	II. It forms monomolecular layers.
III. It takes place at low temperature.	III. It takes place at high temperature.
IV. Adsorption rate increases with increase in pressure of adsorbate.	IV. Adsorption rate decreases with increase in pressure of adsorbate.
V. It is reversible.	V. It is irreversible.
VI. It is not specific.	VI. It is very specific.
VII. Heat of adsorption is 20-40 kJ mol^{-1}.	VII. Heat of adsorption is more than 40 kJ mol^{-1}.

Adsorption of Gases on Solids

The extent of adsorption of a gas on a solid depends upon the following factors:

I. **Nature of the gas:** The easily liquifiable gases are adsorbed to a greater extent than the permanent gases.

II. **Nature of the adsorbent:** Different adsorbents adsorb different gases upto different extent.

III. **Temperature of the system:** Generally, adsorption decreases with the increase in the temperature of the system.

IV. **Pressure of the gas:** Generally, adsorption increases with the increase of the pressure of the gas.

V. Activation of the adsorbent: Solids can be activated by different methods to increase their adsorption powers. This is done by increasing their surface area.

Types of Solutions: Colloidal Solutions

On the basis of particle size of the substance, the solutions may by classiffied into three types:

(a) True solutions

(b) Colloidal solutions and

(c) Suspensions

Emulsions

Emulsions are the colloidal solutions of two immiscible liquids. Emulsions are stabilized by additon of emulsifying agents such as soaps, gum, lyophillic colloids etc. The two types of emulsions are:

I. Oil-in-water type (O/W): Oil is the dispersed phase and water is the dispersion medium. e.g. milk, cream etc.

II. Water-in-oil type (W/O): Water is the dispersed phase and oil is the dispersion medium. Such as butter, cold cream etc.

Applications of Emulsions

I. The concentration of sulphide ore in presence of pine oil by froth floatation process.

II. Digestion of fats in the intestines.

III. Several drugs are available in the form of emulsions.

IV. In cosmetics like creams, lotions and ointments.

V. The cleansing action of soaps and detergents.

VI. Milk is fat-in-oil type emulsion.

Micelles and Cleansing Action of Soaps

Some substances act as electrolytes either at low concentration or at high concentration tend to associate and form aggregated particles which are called as micelles. Such as soaps and detergents. The cleansing action of soap is due to its ability to act as micelle. Soaps are sodium salts of higher fatty acids, such as sodium stearate, $CH_3(CH_2)_{16}COO^-Na^+$. The carboxylic anion COO^- is known as head and it is water soluble. The long chain of alkyl group is called as tail and it is oil soluble. This tail dissolves in the grease deposit (dirt) and form micelle. Which are removed by rinsing with water.

Catalysis

A catalyst is a substance which changes the rate of a chemical reaction without taking share in the reaction. Catalysis is the phenomenon of changing the rate of a reaction with the use of a catalyst. Catalysis are of two types:

I. **Homogeneous catalysis:** When catalyst,reactants and products are in the same phase.

(a) SO_2 is oxidized to SO_3 in presence of nitric oxide.

$$2SO_2(g) + O_2(g) \xrightarrow{NO(g)} 2SO_3(g)$$

(b) Ester hydrolysis:

$$\underset{\text{Ethyl ethanoate}}{CH_3COOC_2H_5(l)} + H_2O(l) \xrightarrow{H^+(aq)} \underset{\text{Ethanoic acid}}{CH_3COOH(l)} + \underset{\text{Ethanol}}{C_2H_5OH(l)}$$

II. **Heterogeneous catalysis:** When catalyst is in different phase to that of reactants. It is a surface phenomenon.

(a) In contact process of manufacture of SO_3 in presence of V_2O_5.

$$\underset{\text{sulphure dioxide}}{2SO_2(g)} + \underset{\text{Oxygen}}{O_2(g)} \xrightarrow{V_2O_5(s)} \underset{\text{Sulphur trioxide}}{2SO_3(g)}$$

(b) Synthesis of CH_3OH from CO and H_2 using ZnO + CuO as catalyst.

(c) In Haber's process of manufacture of NH_3 in presence of Fe catalyst.

$$\underset{\text{Nitrogen}}{N_2(g)} + \underset{\text{Hydrogen}}{3H_2(g)} \xrightarrow{Fe(s)} \underset{\text{Ammonia}}{2NH_3(g)}$$

13

CHEMISTRY OF NON-METALS

Hydrogen : It is the smallest and the lightest element having atomic number one. Its electronic configuration is $1s^1$. It behaves like both alkali and halogen atoms because it has tendency to lose and gain one electron.

Resemblance with Alkali Metals

(i) Electronic configuration: Like alkali metals, hydrogen has 1 electron in its outermost shell.

(ii) Both show tendency to lose valence electron.

(iii) Both possess strong affinity for electronegative elements like oxygen, sulphur, halogens etc.

(iv) Both exhibit oxidation state of +1 and are strong reducing agents.

(v) On electrolysis, both hydrogen and alkali metals are liberated at cathode.

$HCl \rightarrow H^+ + Cl^-$

$2H^+ + 2e^- \rightarrow H_2$ [at cathode]

$NaCl \rightarrow Na^+ + Cl^-$

$Na^+ + e^- \rightarrow Na$ [at cathode]

Resemblance with Halogens

(i) **Electronic configuration :** Both halogens and hydrogen lack one electron to attain configuration of nearest inert gas.

$H \rightarrow 1s^1$ $F \rightarrow 1s^2\,2s^2\,2p^5$. etc.

(ii) **Ionization energy :** Ionization energy of hydrogen is high like that of halogens.

(iii) **Atomic state :** Both form diatomic molecules, like H_2, Cl_2, F_2, Br_2,

(iv) **Oxidation state :** it shows –1 oxidation state like halogens.

(v) **Electrolysis :** In hydrolysis of NaCl, chlorine is liberated at anode while in NaH, hydrogen is liberated at anode.

$2NaCl \rightarrow 2Na + Cl_2$ (at anode)

$2NaH \rightarrow 2Na + H_2$ (at anode)

(vi) **Electronegative character** : Like halogens, hydrogen takes one electron and forms hydride ion.

$H + e^- \rightarrow H^-$

$Cl + e^- \rightarrow Cl^-$

(vii) **Substitution reactions :** Under suitable conditions, hydrogen is replaced by halogens and halogens are replaced by hydrogen.

$$CH_4 + Cl_2 \xrightarrow[\text{Sunlight}]{\text{Diffused}} CH_3Cl + HCl$$

$$CH_3Cl + 2\,[H] \xrightarrow[\text{Zn/HCl}]{\text{Sn/HCl or}} CH_4 + HCl$$

Difference of hydrogen from halogens and alkali metals

(i) Hydrogen is less electropositive than alkali metals but less electronegative than halogens.

(ii) Hydrogen has less tendency to form ions as compared to alkali metals and halogens.

(iii) Size of H ions is much smaller than those of alkali metals ions and halogen ions.

(iv) alkali metal halides have high melting point while pure hydrogen halides have low boiling points.

(v) Hydrogen molecule does not have unshared pair of electrons, while halogen molecules have unshared pair of electrons.

$$H:H \qquad :\ddot{\underset{\cdot\cdot}{Cl}}:\ddot{\underset{\cdot\cdot}{Cl}}:$$

(vi) Oxides of alkali metals are basic, oxides of halogens are acidic while oxide of hydrogen (i.e. water) is neutral.

(vii) Unlike alkali metals and halogens, hydrogen has one electron in extra nuclear part and one proton in nucleus.

Isotopes of Hydrogen

Hydrogen occurs in three isotopic forms : Protium (H), Deuterium (D) and Tritium (T) :

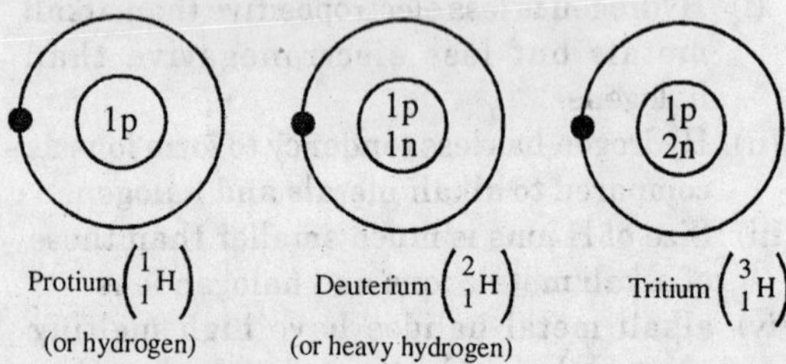

Their chemical properties are the same as their electronic configuration is the same. But physical properties are different due to different atomic masses.

Oxygen : It has atomic numbr 8, electronic configuration is $1s^2\ 2s^2\ 2p^4$. It exists as diatomic molecule. It is paramagnetic in nature because the presence of two unpaired electrons.

Occurrence : Most abundant and occurs both in free as well as in combined form. In free state it is present 21% by volume or 23% by weight in atmosphere. While in combined state it is present in water upto 89%. In earth crust it is present mainly in the form of silicates, aluminates, carbonates and oxides of metals. The balance of oxygen in nature is maintained as it is formed during photosynthesis in plants :

$$\underset{\text{Water}}{xH_2O} + \underset{\text{Carbon dioxide}}{xCO_2} \xrightarrow[\text{Chlorophyll}]{\text{Sunlight}} \underset{\text{Carbohydrate}}{(CH_2O)_x} + \underset{\text{Oxygen}}{xO_2\,(g)}$$

Oxygen occurs in two molecular allotropic forms : dioxygen (O_2) is most stable while ozone (O_3) is less stable. Its three naturally occuring isotopes are ${}^{16}_{8}O$(99.76%), ${}^{17}_{8}O$(0.037%) and ${}^{18}_{8}O$(0.204%).

Uses

(*i*) In the manufacture of nitric acid, sulphuric acid, acetic acid etc.
(*ii*) As an oxidising and bleaching agent.
(*iii*) Used for artificial respiration.
(*iv*) Mixture of liquid oxygen and finely divided charcoal is used as an explosive.
(*v*) As rocket fuel in liquid form.
(*vi*) In oxy-acetylene flames.
(*vii*) In manufacturing steel and in metal fabrication.

Ozone : It is one of the allotropic forms of oxygen. Ozone (O_3) is formed in the upper layer of atomosphere about 20 km above the earth's surface. It does not allow ultraviolet rays to reach the earth. Chlorofluoro carbons deplete ozone layer.

It is prepared by silent electric discharge through cold and dry oxygen in an apparatus called *ozoniser*.

$$3O_2(g) \underset{\text{discharge}}{\overset{\text{Silent electric}}{\rightleftharpoons}} 2O_3(g) \qquad \Delta H = +284 \text{ kJ},$$

Commonly used ozonizers are (*i*) Siemen's ozoniser (*ii*) Brodie's ozoniser

Physical Properties

(*i*) Pure ozone is pale bluish gas with strong pungent odour.

(*ii*) It is slightly soluble in water, readily soluble in oils like turpentine oil, cinnamon oil.

(*iii*) It is heavier than air.

(*iv*) It is highly reactive because of low dissociation energy.

(*v*) It is netural towards litmus.

(*vi*) It is poisonous in nature.

(*vii*) It is a diamagnetic molecule.

Nitrogen : It exists both in combined and free state. In earth's atmosphere it is found upto about 78% by volume. It is present in the form of nitrates like nitre ($NaNO_3$), salt petre (KNO_3) and ammonium salts like ammonium chloride (NH_4Cl) and ammonium sulphate [$(NH_4)_2\ SO_4$]. It is main constituent of animal and plant protein.

Preparation of Nitrogen

(i) From Nitrogen compounds :

(*a*) By heating ammonium nitrite or ammonium dichromate (laboratory preparation) : Sodium nitrite and ammonium chloride are heated:

$$NaNO_2(aq) + NH_4Cl(aq) \xrightarrow{\Delta} NH_4NO_2(aq) + NaCl(aq)$$

$$NH_4NO_2(aq) \xrightarrow{\Delta} N_2(g) + 2H_2O(l)$$

$$\underset{\text{Ammonium dichromate}}{(NH_4)_2Cr_2O_7} \xrightarrow{\Delta} N_2 + Cr_2O_3 + 4H_2O$$

(b) **From ammonia :** By passing ammonia vapours over heated copper oxide (CuO), pure dinitrogen can be obtained.

$$2NH_3 + 3CuO \rightarrow N_2 + 3Cu + 3H_2O$$

(c) From sodium azide :

$$\underset{\text{Sodium azide}}{2NaN_3} \xrightarrow{300°C} 3N_2 + 2Na$$

(d) By the action of nitrous acid on urea :

$$NH_2CONH_2 + 2HNO_2 \rightarrow 2N_2 + CO_2 + 3H_2O$$

(e) By passing vapours of HNO_3 on strongly heated copper :

$$5Cu + 2HNO_3 \rightarrow 5CuO + N_2 + H_2O$$

(ii) **From air (from liquified air):** Liquified air is subjected to fractional distillation. Nitrogen has lower boiling point than oxygen, distills off first. This process is carried out in Claude's apparatus.

Physical properties

(*i*) It is colourless, tasteless and odourless gas.

(*ii*) It is neither combustable nor supports combustion.

(*iii*) It is slightly lighter than air, sparingly soluble in water.

(*iv*) It is non-poisonous in nature but animal do not survive in its atmosphere due to absence of oxygen.

Uses

(*i*) In manufacture of ammonia, nitric acid, nitrides, calcium cyanamide and fertilizers etc.

(*ii*) In filling electric bulbs and gas thermometers due to its low chemical reactivity.

(*iii*) It provids inert atmosphere in metallurgical operations.

(*iv*) Regulates the combustion of oxygen in air.

(*v*) Liquid dinitrogen is used as a refrigerant.

Silicon

It has atomic number 14 and it is the second member of group-14.

It possess certain similarities with carbon, such as

(*a*) Both show covalency of four.

(*b*) Both form hydrides (SiH_4 and CH_4) and halides (SiX_4 and CX_4) of tetrahedral shape.

Silicon also behaves in different manner from carbon due to low electronegativity, availability of d-orbitals and less tendency to form multiple bonds. So, silicon shows following dissimilarities with carbon.

(*a*) Silicon also shows covalency of six unlike carbon which shows covalency of four only.

(*b*) Silicon tetrahalide undergoes hydrolysis while carbon tetrahalide is stable.

(*c*) Carbon dioxide is a gas while silicon dioxide is a solid.

Silicon is the second most abundant element in the earth's crust. It occurs in combined state as silica (SiO_2) in form of sand, flint and quartz and as silicates in the form of mica, felspar ($K_2O.Al_2O_3.6SiO_2$) and kaolinite ($Al_2O_3.2SiO_2.2H_2O$).

Phosphorus : It has atomic number 15 and it is the second member of group-15 after nitrogen. It differs from nitrogen in certain respect due to its large size, low electronegativity and availability of d-orbitals such as nitrogen in sp and sp^2 hybridized state forms stable compounds but compounds of phosphorus in these hybridized states are unstable.

It occurs as phosphates in rocks, in vegetables, in animal bones, blood, teeth and nervous tissues, in egg yolk etc. The important ores of phosphorus are phosphorite [$Ca_3(PO_4)_2$], fluoropatite [$3Ca_3(PO_4)_2 . CaF_2$], hydroxyapatite [$3Ca_3(PO_4)_2 . Ca(OH)_2$], chloropatite [$3Ca_3(PO_4)_2 . CaCl_2$] and bone-ash [$Ca_3 (PO_4)_2$]. Ores of phosphorus are collectively known as phosphate rock.

Phosphorus occurs in three alllotropic forms, white, red and black phosphorus.

White Phosphorus

I. White phosphorus is a white, transparent, soft waxy crystalline solid.

II. It is insoluble in water and alcohol but soluble in carbon disulphide (CS_2) and benzene.

III. It has garlic like odour and is toxic in nature.

IV. It ignites spontaneously in moist air and gives out light (Chemiluminescence). Therefore it is stored under water.

Red Phosphorus

I. It is prepared by heating white phosphorus to about 250° C

$$\text{White P} \xrightarrow[\text{inert atm.}]{240°C} \text{Red P}$$

II. It is odourless and less toxic.

III. It is dark red powder.

IV. It is stable in air and insoluble in organic solvents.

Black Phosphorus

It is prepared by heating white phosphorus under high pressure. It is thermodynamically most stable allotrope.

Fertilizers : These are the chemical substances which are added to the soil to make up for the deficiency of essential elements in plants.

I. **Nitrogeneous fertilizers** : These are the fertilizers which mainly supply nitrogen to the plants. Such as urea (NH_2CONH_2), ammonium sulphate [$(NH_4)_2SO_4$], calcium ammonium nitrate (CAN) [$Ca(NO_3)_2 . NH_4NO_3$].

II. **Phosphatic fertilizers** : These are the fertilizers which mainly supply phosphorus to the plants. Such as triple phosphtate of lime $[Ca(H_2PO_4)_2]$, superphosphate of lime, Thomas slag (double salt of tricalcium phosphate and calcium silicate).

Sulphur : It has atomic number 16 and it is the second member of group-16 after oxygen. Unlike oxygen, sulphur exists in oxidation state of −2, +2, +4 and +6 due to availability of empty d-orbitals.

Sulphur comprises of 0.05% of earth's crust. It occurs in free and combined state both. In free state it occurs in volcanic areas, in limestone cares etc. In combined form as sulphide ores like zinc blende (ZnS), cinnabar (HgS), galena (PbS), copper pyrites ($CuFeS_2$) and iron pyrites (FeS_2) and as sulphate ores like gypsum ($CaSO_4 . 2H_2O$), barytes ($BaSO_4$), epsom salt ($MgSO_4 . 7H_2O$) and celesite ($SrSO_4$).

Sulphur exists in following allotropic forms :

I. **Rhombic sulphur (α) :** It is pale yellow crystalline solid (M.P. 114°C) and most stable of all allotropes of sulphur. It consists of S_8 structural unit packed together into octahedral shape. It is soluble in carbon disulphide and prepared by evaporating sulphur solution in CS_2.

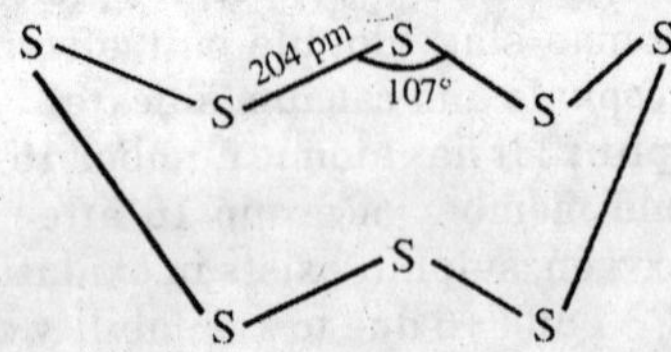

Structure of S_8 molecule

II. Monoclinic (β) : It is bright yellow crystalline solid (M.P. 119°C). It also exists as S_8 molecules but differs from rhombic sulphur in arrangement. It is soluble in carbon disulphide (CS_2) and is prepared by melting and cooling sulphur. It changes into rhombic form above transition temperature (95.6°C).

$$\text{Rhombic sulphur} \underset{}{\overset{95.6^{\circ}C}{\rightleftharpoons}} \text{Monoclinic sulphur}$$

III. Plastic sulphur (γ) : It is soft rubbe like yellowish brown mass (M.P. : not sharp). It has random molecular structure which consists of long chains of sulphur atoms as well as some S_8 rings.It is insoluble in CS_2. It is prepared by pouring boiling sulphur into cold water.

IV. **Colloidal sulphur (δ) :** It can be prepared by passing H_2S through a solution of an oxidising agent in water.

V. **Milk of sulphur :** It can be formed by boiling milk of lime with sulphur and decomposing the product with HCl. It is soluble in CS_2.

Halogens : Fluorine (F), chlorine (Cl), bromine (Br), iodine (I) and astatine (At) are collectively known as halogens. These are all non-metals.

Halogens are very reactive due to high electronegativity and therefore do not exist in free state.

Fluorine : It mainly occurs in fluorspar (CaF_2), cryolite ($AlF_3.3NaF$), fluoropatite [$CaF_2.3Ca_3(PO_4)_2$]. In traces it occurs in sea water, bones, teeth, milk and plants.

Chlorine : Its important sources are sodium chloride (NaCl), potassium chloride (KCl), carnallite (KCl, $MgCl_2.6H_2O$).

Bromine: It is found in sea water as sodium bromide (NaBr), potassium bromide (KBr) and magnesium bromide ($MgBr_2$).

Iodine : Its main sources are sea weeds which contain iodine as alkali metal iodide, caliche contains iodine as sodium iodate ($NaIO_3$), cod liver oil and thyroid glands of animals.

Physical Properties

Fluorine : It is a pale greenish yellow, pungent smelling poisonous gas. It condenses to a pale yellow liquid (B.P. –188°C) and crystallizes to a pale yellow solid (M.P. –233°C). It is heavier than air.

Chlorine : It is a greenish yellow, pungent smelling gas. It dissolves in water and is used as a germicide for water purification.

Bromine : It is reddish brown mobile liquid. It boils at 58.8°C and freezes to a yellowish brown crystalline soild at –7.3°C.

Iodine : It is violet-coloured crystalline solid having metallic lustre.

Noble gases : Group-18 constitues helium (He), neon (Ne), argon (Ar,), krypton (Kr), xenon (Xe) and radon (Rn). These are known as noble gases or inert gases because they have their outermost shell complete and hence do not react normally.

All noble gases occcur in atmosphere except radon which is radioactive. The noble gases occupy 1% by volume of the atmosphere. Helium is second most abundant element in the universe (about 23%) after hydrogen and also occurs in natural gas, in the water of certain springs and in radioactive minerals.

Helium : It is obtained from natural gas. Natural gas is compressed and cooled when nitrogen, hydrocarbons get liquefied. Helium is not liquefied, thus is purified.

Argon, neon, krypton and xenon : Fractional distillation followed by condensation of liquefied air yields neon, argon, krypton and xenon as byproducts.

Radon : Radon is a decay product of radium.

$$\underset{\text{Radium}}{{}^{226}_{88}\text{Ra}} \rightarrow \underset{\text{Radon}}{{}^{222}_{86}\text{Rn}} + \underset{\text{Helium}}{{}^{4}_{2}\text{He}}$$

The most important compunds of noble gases are fluorides of xenon. They are prepared as follows :

$$\underset{(2:1)}{Xe + F_2} \xrightarrow[675\text{ K}]{\text{Sunlight or}} \underset{\text{Xenon difluoride}}{XeF_2}$$

$$\underset{(1:5)}{Xe + 2F_2} \xrightarrow{675\text{ K}} \underset{\text{Xenon tetrafluoride}}{XeF_4}$$

$$\underset{(1:20)}{Xe + 3F_2} \xrightarrow{575-675\text{ K}} \underset{\text{Xenon hexafluoride}}{XeF_6}$$

Uses

Helium:

I. Used in filling metereological balloons and inflating aeroplane tyres.

II. Helium-oxygen mixture is used by divers for artificial respiration and also used in treatment of asthma.

III. To create inert atmosphere and in vacuum tubes, radiotubes, signal lamps etc.

IV. Liquid helium is used to produce extremely low temperature.

Neon: Neon and its combination with other gases is used in discharge tubes called as "Neon signs."

Argon :

I. Used in filling incandescent and fluorescent lamps.

II. For creating inert atmosphere in welding and metallurgical purposes.

Krypton and xenon :

I. Mixture of krypton and xenon is used in electrical flash bulbs for high speed photography.

II. Krypton and xenon are better choice than argon for filling tubes and valves.

Radon : In treacment of cancer and malignats growth.

Some Important Facts

1. **Hydrate :** Many substances combine with water to form crystalline compounds called as hydrates.
2. Polar nature of water molecules is because of charge separation between oxygen and hydrogen atoms.
3. A hydrate crystal contains a definite amount of water, called as water of crystallisation.
4. **Efflorescence :** The phenomenon of crystal becoming anhydrous by losing water of crystallisation is known as efflorescence. e.g. $Na_2CO_3 . 10H_2O$; $CuSO_4 . 5H_2O$.
5. **Deliquescence :** The process in which salt absorbs moisture from the air and then becomes solution, is known as deliquescence.
6. Dehydration : Removal of water form crystals or compounds by heating, is known as dehydration.
7. Heavy water is used as neutron moderator, for production of deuterium, as tracer compound.
8. Calcium hydroxide causes permanent hardness.

9. Soaps are sodium or potassium salts of monovalent long chain fatty acids.
10. Mixture of 3 parts of conc. HCl and 1 part of conc. HNO_3 is called aqua regia.
11. Fuming nitric acid contains dissolved NO_2 in conc. HNO_3. It is brown in colour.
12. Nitrous oxide is known as laughing gas.
13. During generation of steam in boiler, if hard water is used, calcium bicarbonate decomposes into insoluble calcium carbonate, that sticks to inner surface of boiler known as boiler scale. Which causes heat wastage, clogs the tubes of steam and sudden crack in scale cuses boiler to burst.
14. H_2O_2 is a weak acid and forms two series of salts : hydroperoxides and normal peroxides.

$$H_2O_2 + H_2O \rightarrow H_3O^+ + \underset{\text{Hydroperoxide ion}}{HO_2^-}$$

$$HO_2^- + H_2O \rightarrow H_3O^+ + \underset{\text{Peroxide ion}}{O_2^{2-}}$$

15. H_2O_2 acts as a bleaching agent as it releases nascent oxygen that causes oxidation of colouring matter.

16. H_2O_2 should be stored in coloured, wax-lined bottles in presence of stabilizer like H_3PO_4 etc.
17. Boron compounds are electron deficient compounds and therfore act as Lewis acids.
18. Boron forms covalent bonds because of its small atomic size and high ionization energy.
19. Boric acid behaves as a strong monobasic acid in presence of polyhydroxy compounds.
20. Concentrated solution of sodium silicate in water is known as **water glass**.
21. Quartz is the most stable form of silica.
22. White phosphorus is always kept under water.
23. Fluorine cannot be obtained as a product in any chemical reaction
24. Dry chlorine cannot act as bleaching agent. For that moisture is required.
25. There are 8 sulphur atoms in a sulphur molecule.

14

METALLURGY

Various steps of metallurgy are :

1. **Concentration of ore – Crushing**
   (*i*) Levigation – washing
   (*ii*) Magnetic separation
   (*iii*) Froth floatation process
   (*iv*) Leaching
2. **Conversion of metal into metal oxide**
   (*i*) Calcination
   (*ii*) Roasting
3. **Reduction of metal oxides to free metal**
   (*i*) Smelting
   (*ii*) Reduction with CO
   (*iii*) Reduction with highly electropositive metal
   (*iv*) Reduction with H_2
   (*v*) Self reduction
   (*vi*) Reduction by electrolysis

4. Refining or purification of metal

(*i*) Liquation
(*ii*) Distillation
(*iii*) Zone refining
(*iv*) Park's distribution process
(*v*) Cupellation
(*vi*) Poling
(*vii*) Electrolytic refining
(*viii*) Van Arkel process

5. Occurrence of elements

(*i*) ***Elements in atmosphere*:** The atmosphere mainly contains nitrogen (78.09%), oxygen (20.95%) and other geses (about 1%)

(*ii*) ***Elements in sea*:** Sea is the source of elements like Br, I, Ni, Cu, Zn, Sn, Au etc.

(*iii*) ***Elements in earth crust (lithosphere)*:** Elements occur in two states:

(*a*) ***Free or native state*:** Less reactive metals like copper, silver, gold, platinum etc. occur in free state. These metals are generally associated with rocky meterials, sand, clay etc. known as **gangue** or **matrix**.

(*b*) ***Combined state (minerals)***: Reactive metals occur in combined state known as **minerals**. Those minerals from which metal can be profitable extracted are called as **ores**.

6. Classification of ores of elements

(*a*) ***Free or native ores***: Copper, silver, gold and platinum exist in free state.

(*b*) ***Oxide ores***: Bauxite ($Al_2O_3.2H_2O$) of aluminium, haematite (Fe_2O_3) of iron, zincite (ZnO) of zinc, pyrolussite (MnO_2) of manganese are main oxide ores.

(*c*) ***Halide ores***: Carnallite ($KCl. MgCl_2. 6H_2O$) of potassium, rock salt (NaCl) of sodium, cryolite (Na_3AlF_6) of aluminium are main halide ores.

(*d*) ***Carbonate ores***: Calcite ($CaCO_3$) of calcium, dolomite ($MgCO_3. CaCO_3$) of magnesium, malachite [$CuCO_3. Cu(OH)_2$] of copper are main carbonate ores.

(*e*) ***Sulphide ores***: Iron pyrites (FeS_2) of iron, galena (PbS) of lead, copper pyrites (CuS. FeS) of copper, cinnabar (HgS) of mercury, zinc blende (ZnS) of zinc are main sulphide ores.

(*f*) ***Sulphate ores***: Barytes ($BaSO_4$) of barium, alglesite ($PbSO_4$) of lead are main sulphate ores.

(*g*) ***Silicate ores***: Silicon does not occur in free state but is commonly found combined with oxygen, called as silicates. Many elements like, Fe, Mg, K, Na, Ca, Al are found combined with silicates.

Extraction of metals

1. **Concentration of ore:** Concentration of ore provides removal of unwanted substances.

   I. **Crushing of ore:** Big lumps of ore obtained from earth crust are crushed into smaller pieces with the help of jaw crushers and grinders. This process is called as crushing of ore.

II. Removal of impurities from the crushed ore

(a) ***Hand picking*****:** Selected pieces of ores are picked up.

(b) ***Levigation-washing*****:** The crushed ore is washed in a stream of water. The lighter impurities are swept away but heavier ore particles settle down. Iron ores and tin ores are concentrated by this method.

(c) ***Froth floatation process*****:** This process is commonly used for sulphide ores and is based upon different wetting characteristics of ore and gangue particles. Finely powdered ore is mixed with water, pine oil and ethylxanthate or potassium ethyl xanthate in a big tank. The whole mixture is agitate with air. The ore particles wetted with oil come in froth, are taken off but impurities wetted with water settle at the bottom.

(d) ***Magnetic separation*****:** In this method powdered ore is placed over leather belt which moves over two rollers one of which is magnetic. If the crushed ore is passed over magnetic roller, magnetic ore particles are attracted by it and fall below it while impurities fall away from the magnetic roller. Rutile (TiO_2) from apatite, chromite [$Fe(CrO_2)$] from silicious, and wolframite ($FeWO_4$) from cassiterite are separated by this method.

(*e*) ***Leaching***: In this method powdered ore is treated with a suitable chemical reagent that dissolves the ore while impurities remain insoluble in that reagent.

2. Conversion of metal into metal oxide

(a) Calcination: It is the process of heating ore below their melting point in absnece of air to remove volatile impurities like water, CO_2 etc. and organic matter etc.

$$\underset{\text{Lime stone}}{CaCO_3} \rightarrow CaO + CO_2\uparrow$$

$$\underset{\text{Bauxite}}{Al_2O_3.2H_2O} \rightarrow \underset{\text{Alumina}}{Al_2O_3} + 2H_2O\uparrow$$

(b) Roasting: It is the process of heating ore in excess of air so that metals convert into their oxides and water insoluble sulphides to water soluble sulphates.

$$\underset{\text{Pyrite}}{2FeS} + 3O_2 \xrightarrow{\Delta} 2FeO + 2SO_2$$

$$\underset{\text{Cinnabar}}{2HgS} + 3O_2 \xrightarrow{\Delta} 2HgO + 2SO_2$$

$$\underset{\text{Lead sulphide}}{PbS} + 2O_2 \xrightarrow{\Delta} PbSO_2$$

$$\underset{\text{Zinc sulphide}}{ZnS} + 2O_2 \xrightarrow{\Delta} ZnSO_4$$

3. Reduction of metal oxides of free metal

(a) Reduction with carbon (smelting): Reduction of oxides of less electropositive metals like Fe, Pb, Sb, Zn and Cu is carried out by heating them with coal or coke.

$$\underset{\text{Zinc oxide}}{ZnO} + \underset{\text{Coke}}{C} \xrightarrow{\Delta} Zn + CO$$

$$\underset{\text{Lead monoxide}}{PbO} + \underset{\text{Coke}}{C} \xrightarrow{\Delta} Pb + CO$$

Flux is a substance added to remove non-fusible impurities from roasted or calcined ore as fusible substance called as **slag**.

Flux + non-fusible impurity → Fusible slag

Acid flux is used to remove basic impurties like silica (SiO_2)

$$\underset{\substack{\text{Ferric oxide}\\\text{(basic impurity)}}}{Fe_2O_3} + \underset{\substack{\text{Silica}\\\text{(acid flux)}}}{3SiO_2} \rightarrow \underset{\substack{\text{Ferric silicate}\\\text{(slage)}}}{Fe_2(SiO_3)_3}$$

Basic flux is used to remove acidic impurties, like calcium oxide (CaO).

$$\underset{\substack{\text{Phosphours}\\\text{pentoxide}\\\text{(Acidic impurity)}}}{P_2O_5} + \underset{\substack{\text{Calcium}\\\text{oxide}\\\text{(basic flux)}}}{3CaO} \rightarrow \underset{\substack{\text{Calcium}\\\text{phosphate}\\\text{(slag)}}}{Ca_3(PO_4)_2}$$

(*b*) **Reduction with carbon monoxide:** CO produced by heating coke in limited supply of oxygen, is also used as reducing agent.

$$Fe_2O_3 + 2CO \rightarrow 2Fe + 3CO_2$$

Ferric oxide

$$PbO + CO \rightarrow Pb + CO_2$$

Lead monoxide, Carbon monoxide, Lead, Carbon doxide

(c) **Reduction with highly electropositive metal:** Some metal oxides which are not reduced by carbon, such as titanium chloride ($TiCl_4$), chromium trioxide (Cr_2O_3), manganese oxide (Mn_3O_4) are reduced by using Al, Mg etc.

$$3Mn_3O_4 + 8Al \rightarrow 9Mn + 4Al_2O_3 + Heat$$

Manganese oxide, Aluminum oxide

$$Cr_2O_3 + 2Al \rightarrow 2Cr + Al_2O_3 + Heat$$

Chromium trioxide, Aluminium oxide

$$TiCl_4 + 2Mg \xrightarrow{\Delta} Ti + 3MgCl_2$$

Titanium trioxide, titanium, Magnesium chloride

(d) **Reduction with hydrogen:**

$$WO_3 + 3H_2 \xrightarrow{\Delta} W + 3H_2O$$

Tungsten trioxide, Tungsten

$$\underset{\text{Molybdenum trioxide}}{MoO_3} + 3H_2 \rightarrow \underset{\text{Molybdenum}}{Mo} + 3H_2O$$

(e) Self reduction: Sulphide ores of less electropositive metals such as Hg, Cu, Pb, Sb etc. undergo self reduction.

$$\underset{\text{Mercury (II) oxide}}{2HgO} + \underset{\text{Mercury (II) sulphide}}{HgS} \rightarrow 3Hg + SO_2$$

$$\underset{\text{Cinabar}}{2HgS} + 3O_2 \rightarrow \underset{\text{Mercury (II) oxide}}{2HgO} + 2SO_2$$

$$2PbO + PbS \rightarrow 3Pb + SO_2$$

$$2PbS + 3O_2 \rightarrow 2PbO + 2SO_2$$

(f) Reduction by electrolysis: Alkali and alkaline metals are extracted by this method. Such as sodium metals is obtained by the electrolysis of fused sodium chloride.

$$NaCl \rightarrow Na^+ + Cl^-$$

At cathode: $Na^+ + e^- \rightarrow Na$

At anode : $Cl^- - e^- \rightarrow Cl$

$Cl^{\cdot} + Cl \rightarrow Cl_2$

Some Important Facts

I. Alkali and alkaline earth metals being good reducing agents cannot be prepared by reduction of their oxides.

II. Alkali metals are highly electropositive, so also cannot be prepared by displacing them from their salts.

III. Alkali metals dissolve in ammonia to form blue solution which is good conductor of electricity due to presence of solvated electrons.

IV. KOH is a better absorber of CO_2 than NaOH.

V. Sodium hydroxide is called as caustic soda because it breaks down the protein of skin to a pasty mass.

VI. Gypsum is added to cement to prolong its setting time.

VII. Magnesia cement is the saturated solution of magnesium chloride and magnesium oxide ($MgCl_2 . 5MgO . xH_2O$).

VIII. Thomas slag formed in Bessemer process is used as fertilizer.

IX. Zinc atom is larger than copper atom, because it has fully filled d-orbitals.

X. Hardness of steel increases with increase in carbon content.

CHEMISTRY OF METALS

Sodium (Na) and Potassium (K)

Sodium and potassium are very reactive elements and hence do not occur in free state. In combined state they are quite abundant in earth crust.

Ores of sodium : Rock salt (NaCl), salt petre ($NaNO_3$), sodium sulphate ($Na_2SO_4.10H_2O$), borax ($Na_2B_4O_7.10H_2O$), sodium carbonate (Na_2CO_3), albite ($NaAlSi_3O_8$) etc.

Ores of potassium : Carnallite ($KCl.MgCl_2.6H_2O$) sylvinite (KCl.NaCl), felspar ($K_2O.Al_2O_3.6SiO_2$) etc.

Extraction of Sodium

Sodium is prepared by electrolysis of 40% sodium chloride and 60% calcium chloride in Down's cell. On electrolysis, sodium is liberated at anode and collected in an inverted trough. Down's cell is made up of steel, lined with heat resistant bricks. The graphite anode and iron cathode are separated by steel metal gauze to prevent mixing sodium and chlorine.

$$NaCl \rightarrow Na^+ + Cl^-$$

At cathode: $Na^+ + e^- \rightarrow Na$

At anode: $2Cl^- - 2e^- \rightarrow Cl_2$

Calcium chloride is added to lower the melting point of sodium chloride from (803°C to about 600°C). The main reasons for lowering the temperature are :

(i) It is difficult to attain and maintain 803°C temperature.

(ii) Formation of metallic fog is prevented.

(iii) Collection of sodium metal becomes easier as sodium vapours ignite in air at high temperature.

(iv) At low temperature, sodium and chlorine do not corrode the electrolytic cell.

Extraction of potassium : It connot be obtained by electrolysis of potassium chloride because potassium is easily soluble in potassium chloride and does not float on the top of the cell.

$$KOH \rightleftharpoons K^+ + OH^-$$

At cathode: $K^+ + e^- \rightarrow K$

At anode: $4OH^- \rightarrow 2H_2O + O_2 + 4e^-$

Potassium metal can be also obtained by reduction of potassium chloride with sodium metal at 900°C.

$$KCl + Na \xrightarrow{900°C} NaCl + K$$

Sodium and potassium are soft silvery white metals and can be easily cut with a knife. Their surface gets tarnished if exposed to air due to formation of oxides or carbonates. These metals are stored under kerosene.

Uses of Sodium

I. It is used to manufacture sodium peroxide, sodium cyanide, sodamide etc.

II. As a coolant in nuclear reactors because it has low B.P. and high thermal conductivity.

III. In sodium vapour lamps.

IV. Sodium amalgam is used in synthesis of organic compounds as a reducing agent.

V. Sodium-lead alloy is used in the preparation of lead tetraethyl and antiknock agent.

Uses of Potassium

I. Mixture of potassium and caesium is used in photoelectric cells.

II. Potassium bromide is used in photography.
III. As a reducing and laboratory reagent.
IV. Potassium chloride, potassium sulphate and potassium nitrate are used as fertilizers.
V. Amalgam of potassium is used in preparation of organic compounds.

Magnesium (Mg)

Its minerals are carnellite ($KCl . MgCl_2 . 6H_2O$), magnesite ($MgCO_3$), dolomite ($MgCO_3 . CaCO_3$), asbestos [$CaMg_3(SiO_3)_4$], epsom salt ($MgSO_4 . 7H_2O$) and kieserite ($MgSO_4 . 2H_2O$).

Extraction

It is obtained by Dow's sea water process.

Isolation of $MgCl_2.6H_2O$: Calcium hydroxide when added to sea water precipitates magnesium hydroxide from it. Magnesium hydroxide on treatment with hydrochloric acid gives magnesium chloride.

$$MgCl_2 + Ca(OH)_2 \xrightarrow{-CaCl_2} Mg(OH)_2\downarrow \xrightarrow{\text{Dil. HCl}} MgCl_2.6H_2O$$

Preparation of anhydrous $MgCl_2$: Anhydrous magnesium chloride cannot be prepared by heating as it is hydrolysed by its water of crystallization and forms magnesium oxide.

$$MgCl_2.6H_2O \xrightarrow{\Delta} MgO + 2HCl + 5H_2O$$

Anhydrous magnesium chloride is obtained by passing current of dry hydrochloric acid through it.

$$MgCl_2.6H_2O + HCl \rightarrow MgCl_2 + 6H_2O + HCl$$

Electrolysis of anhydrous magnesium chloride : Electrolysis of fused mixture of magnesium chloride and soidum chloride is carried out in steel tank at 700°C.

$$MgCl_2 \rightarrow Mg^{2+} + 2Cl^-$$

At cathode: $Mg^{2+} + 2e^- \rightarrow Mg$

At anode: $2Cl^- - 2e^- \rightarrow Cl_2$

Sodium chloride is added to magnesium chloride to reduce its melting point and to increase its electrical conductivity.

Physical Properties

(*i*) It is a silvery white hard metal (M.P. 650°C)

(*ii*) It is one of the lightest metal.

(*iii*) It is good conductor of heat and electricity.

(*iv*) It is sold in the form of ribbons or wires as it is maleable and ductile.

Uses

(i) It is used in flashlight photography.

(ii) It is used as a reducing agent in preparation of boron and silicon from their oxides.

(iii) It is used as a de-oxidizer in making brass.

(iv) It is used in manufacture of alloys like magnelium (95% Al and 5% Mg) used for making blances and duralium (95% Al, 4% Cu, 0.5% Mg, 0.5% Mn) used in construction of aircrafts etc.

(v) It is used for preparing Grignard reagent which is widely used in organic chemistry.

Calcium (Ca)

Occurrence : *Minerals of calcium :* Limestone ($CaCO_3$), fluorspar (CaF_2), gypsum ($CaSO_4.2H_2O$), fluorapatite [$3Ca_3.(PO_4)_2.CaF_2$] and phosphorite [$Ca_3(PO_4)_2$].

Extraction of calcium : It is extracted by the electrolysis of a fused mixture of calcium chloride with 16% calcium fluoride.

$$CaCl_2 \rightarrow Ca^{2+} + 2Cl^-$$

At cathode : $Ca^{2+} + 2e^- \rightarrow Ca$

At anode : $2Cl^- - 2e^- \rightarrow Cl_2$

Physical Properties

(*i*) It is a silvery white, lusturous metal. If exposed to air, its surface gets tarnished due to oxidation.

(*ii*) It is good conductor of heat and electricity.

(*iii*) It is malleable and hard like tin.

Uses

I. In iron and steel industry as a deoxidizer for cast iron.

II. As a reducing agent.

III. As a dehydrating agent for the preparation of absolute alcohol.

IV. As a scavenger for phosphorus, oxygen and sulphur.

V. For making alloys with Al for bearings.

VI. For removing last traces of nitrogen from inert gases and vacuum tubes.

Aluminium (Al)

Occurrence : Its important ores are as follows :

Oxides : Corrundum (Al_2O_3), diaspore ($Al_2O_3 . H_2O$), bauxite ($Al_2O_3 . 2H_2O$) and gibbsite ($Al_2O_3 . 3H_2O$).

Fluoride : Cryolite (Na_3AlF_6)

Sulphate : Alunite or alumstone [$K_2SO_4 . Al_2(SO_4)_3 . 4Al(OH)_3$].

Silicates : Feldspar ($KAlSi_3O_8$), mica [$KAlSi_3O_{10}(OH)_2$].

Extraction : Aluminium is extracted from bauxite ore by following method :

(a) *Baeyer's process for removal of ferric oxide* : Sodium hydroxide is added to the ore, ferric oxide remains undissolved but Al dissolves and forms sodium aluminate. Aluminium hydroxide is precipitated from sodium aluminate by treating the solution with alumina. At last the precipitate of aluninium hydroxide is filtered, washed, dried and heated.

(b) *Serpeck's process for removal of silica* : It involves heating of powdered ore with coke at 1800°C when silica is reduced to silicon and vapories off.

Electrolytic Reduction of Pure Alumina

Alumina is a bad conductor of electricity and it has high melting point, is electrolysed by dissolving in cryolite (Na_3AlF_6) and adding fluorospar (CaF_2) to it. The electrolysis is carried out in iron tank lined inside with gas carbon that serves as cathode and carbon rods dipped in fused electrolyte serve as anode. On passing current through it, Al is obtained at cathode.

$$Al_2O_3 \rightleftharpoons Al^{3+} + AlO_3^{-3}$$

At cathode: $Al^{3+} + 3e^- \rightarrow Al$

At anode: $4AlO_3^{3-} \rightarrow 2Al_2O_3 + 3O_2 + 12e^-$

Physical Properties

(*a*) Pure aluminium is silvery white lusturous metal (M.P. 932 K and B.P. 2743K).

(*b*) It is mealleable, ductile and is a good conductor of heat and electricity.

Uses

I. Aluminium is malleable and can be beaten into thin foils, so it is used for packing medicines, food articles, cigarettes etc.

II. Being a good conductor of electricity, it is used in making transmission wires.

III. Used for making silvery paints.

IV. Used in thermite welding.

V. Because of good thermal conductivity, it is used for making cooking utensils.

VI. For making various alloys.

VII. Used in making aircrafts and body parts of automobiles.

VIII. Aluminium amalgum is used as reducing agent.

Iron (Fe)

It is the second most abundant metal in the earth crust after aluminium. Ores of iron are : Haematite (Fe_2O_3), Limonite ($Fe_2O_3.2H_2O$), Magnetite (Fe_3O_4), Siderite ($FeCO_3$), Iron pyrities (FeS_2) and chalcopyrites ($CuFeS_2$).

Extraction of iron : It is extracted from haematite by following process :

Concentration of ore : The crushed ore is washed with water to remove sand, clay etc.

Calcination and roasting : The concentrated ore is roasted in excess of air.

(*i*) To remove moisture, phosphorus, carbon dioxide and arsenic as their volatile oxides.

(*ii*) To oxidize ferrous oxide to ferric oxide.

$$\underset{\text{Ferrous oxide}}{4FeO} + \underset{\text{Oxygen}}{O_2} \rightarrow \underset{\text{Ferric oxide}}{2Fe_2O_3}$$

(*iii*) To make ore porous then making it suitable for easy reduction to iron.

Smelting : Blast furnace is charged with lime stone, $CaCO_3$ (flux), coke (a reductant) and haematite. Then hot air blast is forced from the bottom of the furnace. Reactions taking place in its different zones are :

I. Combustion zone (1500°C – 2000°C).

$$\underset{\text{Coke}}{C} + O_2 \rightarrow CO_2 \quad \Delta H = -400 kJ$$

II. **Absorption zone (800°C – 1000°C) :** Carbon dioxide rising up form combustion zone is reduced to carbon monoxide.

$$CO_2 + C \rightarrow 2CO \quad \Delta H = +160 kJ$$

III. **Reduction zone (400°C – 700°C) :** In this zone charge moving down meets carbon monoxide.

$$Fe_2O_3 + 3CO \rightarrow 2Fe + 3CO_2$$

$$Fe_3O_4 + CO \rightarrow 3FeO + CO_2$$

$$FeO + CO \rightarrow Fe + CO_2$$

IV. **Slag formation zone :** In this zone limestone decomposes to give calcium oxide then with silica (impurity) forms fusible slage.

$$CaCO_3 \rightarrow CaO + CO_2$$

$$CaO + \underset{\text{Silica}}{SiO_2} \rightarrow \underset{\text{Calcium silicate (slag)}}{CaSiO_3}$$

V. **Fusion zone (1500°C)** : It is the lowest part of the furnace form where molten iron is tapped off.

Forms of Iron

(*a*) **Cast iron or pig iron :** It is most impure form of iron. It contains about 3–5% of carbon and other impurities like Si, P, Mn and S due to which it is very hard and brittle.The two types of cast iron are white cast iron and grey cast iron. Cast iron is used for casting articles as it expands on solidification. It cannot be welded and magnetized permanetly.

(*b*) **Wrought iron :** It is the most pure form of iron. It is extermely though, malleable, ductile and resistant towards rusting and corrosion. It contains least amount of carbon (about) (0.2–0.5%).

Preparation of wrought iron : It is prepared by heating cast iron with haematite in reverberatory furnace. Oxides of carbon and sulphur escape out as gases but oxides of manganese, phosphorus and silicon form slag.

$$\underset{\text{Haematite}}{Fe_2O_3} + 3C \rightarrow 2Fe + 3CO\uparrow$$

$$MnO + SiO_2 \rightarrow MnSiO_3$$

$$\underset{\text{Haematite}}{2Fe_2O_3} + P_4O_{10} \rightarrow \underset{\text{(slag)}}{4FePO_4}$$

Uses of Wrought Iron : It is used to make hourse shoes, iron nails, magnets, chains, fire-bars and anchors etc. as it can withstand sudden stress because presence of slag between its different layers.

(*c*) **Steel :** Steel is the most useful iron which contains 0.2–2% of carbon. Various types of steel are :

(*i*) **Mild steel :** it contains about 0.1–0.4% carbon.

(*ii*) **Hard steel :** It contains carbon upto 1.5%.

(*iii*) **Alloy steel :** It has small amounts of nickel, cobalt, chromium, tungsten, molybdenum, manganese etc. are added to obtain desired properties are known as alloy steel. e.g. stainless steel (Cr 18%, Ni 8%) does not rust or corrode and is used for making utensils, shaving blades etc. Alnico (Al 12%,

Ni 20%) that is higlly magnetic and is used for making permanent magnets.

Copper (Cu) : It exists in earth crust in free state (due to low reactivity) as well as in combined state. Its main ores are cuprite (Cu_2O), copper pyrites ($CuFeS_2$), copper glance (Cu_2S), indigo copper (CuS) malachite [$Cu(OH)_2 . CuCO_3$] and azurite [$Cu(OH)_2 . 2CuCO_3$].

Extraction : It is extracted form its sulphide ore copper pyrites ($CuFeS_2$).

Concentration of ore : Copper pyrites ore is concentrated by froth floatation process.

Roasting : Roasting is carried out by heating concentrated ore in presence of air to remove volatile impurties in the form of their oxides.

$$\underset{\text{Copper pyrites}}{2CuFeS_2} + O_2 \rightarrow Cu_2S + 2FeS + SO_2$$

$$2Cu_2S + 3O_2 \rightarrow 2Cu_2O + 2SO_2$$

$$2FeS + 3O_2 \rightarrow 2FeO + 2SO_2$$

$$S + O_2 \rightarrow SO_2 \uparrow$$

$$4As + 5O_2 \rightarrow 2As_2O_5 \uparrow$$

$$4P + 5O_2 \rightarrow 2P_2O_5 \uparrow$$

Smelting : The roasted ore is heated with coke (to produce heat) and sand (to form fusible slag) in the presence of air in blast furnace.

$$2FeS + 3O_2 \rightarrow 2FeO + 2SO_2$$

$$FeO + SiO_2 \rightarrow FeSiO_3$$

$$Cu_2O + FeS \rightarrow Cu_2S + FeO$$

Copper sulphide still contains some ferrous sulphide. It is known as matte.

Bessemerization : A blast of hot air and sand is blown through a Bessemer converter charged with molten matte. Ferrous sulphide present in matte is oxidized to ferrous oxide and forms fusible slag with silica that is drained off. Cuprous sulphide is partially oxidized to cuprous oxide and reacts with cuprous sulphide to form copper.

$$2Cu_2O + Cu_2S \rightarrow 6Cu + SO_2$$

Volatile impurities come out during cooling of copper and large blisters are formed on the surface. Copper then obtained is known as blister copper.

Refining of copper : Refining of copper is done either by poling or by electrolytic refining. In electrolytic refining of copper, acidified copper sulphate is used as electrolyte, impure copper metal serves as anode and a plate of pure copper is made cathode.

At cathode : $Cu^{2+} + 2e^- \rightarrow Cu$ (Pure copper)

At anode : $Cu \rightarrow Cu^{2+} + 2e^-$ (Impure cupper)

Uses

I. Being a good conductor of electricity, it is used in the manufacture of electrical wires, cable etc.

II. As it is good conductor of heat and resistant towards steam, it is used for making household untensils, boilers etc.

III. Wide range of alloys of copper are used for various purposes.

IV. For electroplating and electrotyping.

V. Used in making ornaments and coins.

Silver (Ag) : It exists in native state as well as in combined state as horn silver (AgCl), silver glance or argentite (Ag_2S), ruby silver or pyrargyrite ($Ag_2S . SbS_3$).

Extraction : It is extracted from silver glance by cyanide process.

Concentration : Crushed ore is concentrated by froth floatation process.

Treatment with sodium cyanide : Silver sulphide present in concentrated ore forms soluble complex with sodium cyanide in presence of air, but impurities are filtered off.

$$\underset{\text{Silver sulphide}}{Ag_2S} + 4NaCN \rightarrow \underset{\text{Sodium argento cyanide}}{2Na[Ag(CN)_2]} + Na_2S$$

$$4Ag + 8NaCN + O_2 + 2H_2O \rightarrow 4Na[Ag(CN)_2] + 4NaOH$$

$$AgCl + 2NaCN \rightarrow Na[Ag(CN)_2] + NaCl$$

$$4Na_2S + 5O_2 + 2H_2O \rightarrow 2Na_2SO_4 + 4NaOH + 2S$$

Recovery of silver : It is recovered by adding scrap zinc to the cyanide complex soluton.

$$2[Ag(CN)_2]^- + Zn \rightarrow [Zn(CN)_4]^{2-} + 2Ag$$

Refining : It is refined by electrolysis of a solution of silver nitrate and 10% nitric acid having impure silver as anode and pure silver as cathode.

Uses

I. In making coins, oranaments, silver wares etc.

II. In high capactiy Ag-Zn and Ag-Cd batteries.

III. In silvering of mirrors and in photography.

IV. In medicines and silver amalgam in filling teeth.

Gold (Au) : It occurs native in alluvial deposits, in quartz veins and in lead or copper sulphide ores. In combined state it occurs as tellurides, calaverite ($AuTe_2$) and sylvanite ($AuAgTe_2$).

Extraction

I. **Amalgamation process :** Gold rock are crushed and its amalgam is prepared with mercury. On distillation of amalgam, gold is obtained.

II. **MacArthur-Forrest cyanide process :**

Concentration of ore : The crushed ore is concentrated by froth-floatation process.

Treatment with potassium cyanide : Here, the concentrated ore is leached with potassium cyanide in presence of air.

$$4Au + 8KCN + 2H_2O + O_2 \rightarrow \underset{\text{Potassium dicyanoaurate (I)}}{4K[Au(CN)_2]} + 4KOH$$

Recovery : Gold precipitates down if the solution is treated with more electropositive metal than gold like zinc.

$$2K\,[Au(CN)_2] + Zn \rightarrow \underset{\text{Potassium tetra cyanozincate (II)}}{K_2\,[Zn(CN)_4]} + 2Au$$

Refining : Impure gold thus obtained is refinet by one of the following methods :

(*a*) **Parting :** Impure gold is boiled with concentrated sulphuric acid or concentrated nitric acid. Silver and copper are soluble in acid thus they go into the solution, while gold is left behind.

$$2Ag + 2H_2SO_4 \rightarrow Ag_2SO_4 + SO_2 + 2H_2O$$

$$Cu + 2H_2SO_4 \rightarrow CuSO_4 + SO_2 + 2H_2O$$

(b) Electrolytic refining : Electrolytic refining of gold is carried out by using gold (III) chloride acidified with 10-12% hydrochloric acid as electrolyte, impure gold as anode and pure gold as cathode. Thus highly pure gold is obtained by this method.

Properties

I. It is yellow lusturous, malleable, ductile and noble metal.

II. It is resistant towards attack of air, water, acids or alkalies.

III. Reaction with aqua regia (3 parts HCl + 1 part HNO_3). It dissolves in aqua regia.

$$\underbrace{3HCl + HNO_3}_{\text{Aqua regia}} \rightarrow \underset{\text{Nitrosyl chloride}}{NOCl} + 2H_2O + Cl_2$$

$$2Au + 3Cl_2 \rightarrow \underset{\text{Gold (III) Chloride}}{2AuCl_3}$$

Uses

I. It is used in the manufacture of jewellery and coins.

II. For gold plating.

Zinc (Zn) : It occurs in combined state and its important ores are zincite (ZnO), zinc ferrite ($ZnO.Fe_2O_3$), zinc spinel ($ZnO_2 . AlO_3$), zinc blende (ZnS), calamine ($ZnCO_3$) and willemite (Zn_2SiO_4).

Extraction : It is extracted form zinc blende.

Concentration of ore : Finely powdered ore is concentrated by froth floatation process.

Roasting : The concentrated ore is heated in excess of air at 1200 K to convert zinc sulphide and zinc sulphate to zinc oxide.

$$2ZnS + 3O_2 \xrightarrow{\Delta} 2ZnO + 2SO_2$$

$$ZnS + 2O_2 \xrightarrow{\Delta} ZnSO_4$$

$$2ZnSO_4 \xrightarrow{\Delta} 2ZnO + 2SO_2 + O_2$$

When calamine is used, it decomposes to zinc oxide and carbon dioxide.

$$\underset{\text{(Calamine)}}{ZnCO_3} \xrightarrow{\Delta} ZnO + CO_2$$

Reduction : Zinc oxide is reduced by coke in a blast furnace.

$$ZnO + C \rightarrow Zn + CO$$

Electrolytic refining : Zinc is refined by electrolytic method by using zinc sulphate with dilute sulphuric acid as electrolyte, impure zinc as anode and pure zinc as cathode.

At cathode: $Zn^{2+} + 2e^- \rightarrow Zn$

At anode: $2H_2O \rightarrow 4H^+ + O_2 + 4e^-$

Uses

I. It is used in galvanising iron and making alloys like brass and german silver.

II. In extraction of silver and gold by cyanide process.

III. As a reducing agent.

IV. In making electric batteries.

Mercury (Hg) : The only important ore of mercury is cinnabar (HgS).

Extraction

Concentration : The ore is concentrated by froth floatation process.

Simultaneous roasting and distillation : The concentrated ore is roasted in excess of air to oxidise mercury (II), sulphide (cinnabar) to mercury (II) oxide.

$$2HgS + 3O_2 \rightarrow 2HgO + 2SO_2$$

Further, mercury (II) oxide decomposes to produce mercury vapours that are condensed.

$$2HgO \rightarrow 2Hg + O_2$$

Refining : Impure mercury thus obtained contains copper, zinc, lead and bismuth. These are removed by following methods.

I. Filteration

II. Treatment with dilute nitric acid. It involves dissolution of all metals while mercury remains unaffected.

III. Distillation under reduced pressure.

Uses

I. It has high boiling point and low freezing point, therefore used in making barometers and thermometers.

II. In the manufacture of caustic soda and as a catalyst in many chemical reactions.

III. Silver and gold amalgam are used for dental filling.

IV. In the preparation of calomel which is used to prepare standard electrode.

V. In silvering of mirrors, mercury vapour lamps and in the extraction of silver and gold.

Tin (Sn) : The important ore of tin is cassiterite (SnO_2).

Extraction

Concentration : The finely powdered ore is concentrated by gravity process.

Roasting : The ore is roasted in a current of air if impurities like sulphur and arsenic escape as volatile oxides. Iron and copper pyrites are converted into their oxides and sulphates.

$$4As + 3O_2 \rightarrow 2As_2O_3 \uparrow$$

$$S + O_2 \rightarrow SO_2 \uparrow$$

Electromagnetic separation : Wolframite (ore of tungsten) and iron oxide are separated from cassiterite by magnetic separation.

Washing : The concentrated ore is treated with water to dissolve out ferrous sulphate and copper sulphate, but lighter impurities are washed off. Thus heavier tin sinks to the bottom and it is referred as **black tin**.

Smelting : Black tin is heated with coke to reduce tin oxide to tin metal and with limestone that acts as flux to remove impurities.

$$\underset{\text{Cassiterite}}{SnO_2} + \underset{\text{Coke}}{2C} \rightarrow Sn + 2CO$$

$$\underset{\text{(flux)}}{CaCO_3} \rightarrow CaO + CO_2$$

$$CaO + \underset{\text{(impurity)}}{SiO_2} \rightarrow \underset{\text{(fusible slag)}}{CaSiO_3}$$

Refining : Impure tin metal is refined by following methods :

I. Liquation.

II. Electrolytic refining containing tin sulphate with fluorosilicic acid (H_2SiF_6) and sulphuric acid as the electrolyte, impure tin as anode and pure tin sheet as cathode.

$$PbS + PbSO_4 \xrightarrow{\Delta} 2Pb + 2SO_2$$

Uses

I. For tin plating iron, for wrapping foils.

II. Tin amalgam is used for making mirrors.

III. It is used for making alloys.

Lead (Pb) : Its important ores are galena (Pbs), anglesite ($PbSO_4$), cerussite ($PbCO_3$) and lanarkite ($PbO.PbSO_4$).

Extraction : It is extracted from galena by following ways :

Concentration : The powdered ore is concentrated by froth-floatation process.

Reduction : The concentrated ore is reduced to the metal by following process :

I. **Air reduction process :** The concentrated ore is roasted in the reverberatory furnace in a limited supply of air.

$$2PbS + 3O_2 \rightarrow 2PbO + 2SO_2$$

$$PbS + 2O_2 \rightarrow PbSO_4$$

Smelting : Then the temperature is raised and air supply is cut off.

$$PbS + 2PbO \xrightarrow{\Delta} 3Pb + SO_2$$

$$PbS + PbSO_4 \xrightarrow{\Delta} 2Pb + 2SO_2$$

II. **Carbon reduction process :** The concentrated ore is mixed with lime (flux) and roasted in excess of air.

$$2PbS + 3O_2 \rightarrow 2PbO + 2SO_2$$

$$S + O_2 \rightarrow SO_2 \uparrow$$

$$4As + 3O_2 \rightarrow 2As_2O_3 \uparrow$$

Smelting : Coke is added to roasted ore and smelted.

$$PbO + \underset{\text{Coke}}{C} \rightarrow Pb + CO$$

$$PbO + CO \rightarrow Pb + CO_2$$

Electrolytic refining : Impure Pb is refined by taking a solution of lead fluoro-silicate

($PbSiF_6$) with 8-12% of hydrofluoro silicic acid (H_2SiF_6) as electrolyte, impure lead as anode and pure lead as cathode.

Uses

I. To prepare lead tetraethyl, $(C_2H_5)_4Pb$ which is used as an antiknocking agent in petrol.

II. In making useful alloys such as solder, type metal, pewter etc.

III. In the manufacture of sulphuric acid for making chambers in lead chamber process.

IV. For making telegraph and telephone wires buried under earth.

V. For making lead pigments, lead pipes, lead accumulators and bullets etc.

IMPORTANT ORES

	Ore	Formulae	Ore of metal
1.	Alumina	Al_2O_3	Aluminium
2.	Alunite	$K_2SO_4 \cdot Al_2(SO_4)_3 \cdot 4Al(OH)_3$	Aluminium
3.	Anhydrite	$CaSO_4$	Calcium
4.	Argentite	Ag_2S	Silver
5.	Azurite	$2CuCO_3 \cdot Cu(OH)_2$	Copper

	Ore	Formulae	Ore of metal
6.	Brine	NaCl (Solution)	Sodium
7.	Borax	$Na_2B_4O_7.10H_2O$	Boron
8.	Bauxite	$Al_2O_3.2H_2O$	Aluminium
9.	Calaverite	$AuTe_2$	Gold
10.	Chile Saltpeter	$NaNO_3$	Sodium
11.	Cinnabar	HgS	Mercury
12.	Chlorapatite	$Ca_5(PO_4)_3Cl$	Calcium
13.	Calcia	CaO	Calcium
14.	Chalk, Marble, Aragonite, Calcite, Iceland spar, Limestone	$CaCO_3$	Calcium
15.	Carnallite	$KCl.MgCl_2.6H_2O$	Magnesium
16.	Calamine	$ZnCO_3$	Zinc
17.	Cassiterite	SnO_2	Tin
18.	Copper Pyrites, Chalcopyrite	$CuFeS_2$	Copper
19.	Copper glance, Chalcocite	Cu_2S	Copper
20.	Cuprite	Cu_2O	Copper
21.	Clay, Kaolin, Chinaclay,Mica, Feldspar	Aluminosilicates	Aluminium
22.	Corundum emery	Al_2O_3	Aluminium
23.	Cryollite	Na_3AlF_6	Aluminium
24.	Diaspore	$Al_2O_3.H_2O$	Aluminium

	Ore	Formulae	Ore of metal
25.	Dolomite	$MgCO_3.CaCO_3$	Magnesium
26.	Epsom salt	$MgSO_4.7H_2O$	Magnesium
27.	Fluorapatite	$3Ca_3(PO_4)_2.CaF_2$	Calcium
28.	Fluorspar	CaF_2	Calcium
29.	Greenockite	CdS	Cadmium
30.	Gypsum	$CaSO_4.2H_2O$	Calcium
31.	Galena	PbS	Lead
32.	Heavy spar	$BaSO_4$	Barium
33.	Horn silver, chloragyrite	$AgCl$	Silver
34.	Haematite (red)	Fe_2O_3	Iron
35.	Nitre	KNO_3	Potassium
36.	Magnesite	$MgCO_3$	Magnesium
37.	Malachite	$CuCO_3.Cu(OH)_2$	Copper
38.	Magnetite	Fe_3O_4	Iron
39.	Pyragyrite	Ag_3SbS_3	Silver
40.	Pyrolusite	MnO_2	Manganese
41.	Rock salt	$NaCl$	Sodium
42.	Schonite	$K_2SO_4.MgSO_4.6H_2O$	Potassium
43.	Sylvanite	(Ag, Au) Te	Gold
44.	Sylvine	KCl	Potassium
45.	Trona, Natrona	Na_2CO_3	Sodium
46.	Whitherite	$BaCO_3$	Barium
47.	Washing soda	$Na_2CO_3.10H_2O$	Sodium
48.	Zincite	ZnO	Zinc
49.	Zinc blends	ZnS	Zinc

16

SOME IMPORTANT ALLOYS, THEIR COMPOSITONS & USES

1. Alnico	63% Fe, 12%Al, 20%Ni, 5%Co (For making permanent magnets)
2. Alpan	Al + Si
3. Almalgam	Hg + any other metal
4. Bell metal	80%Cu, 20%Sn (bells, utencils, idols, coins etc.)
5. Bearing metal	82%Sn, 14%Sb, 4%Cu
6. Birmabright	5%Mg, 95%Al
7. Brass	Cu + Zn (Household utencils)
8. Britannia metal	93%Sn, 5%Sb, 2%Cu
9. Bronze	75% to 90%Cu, 25% to 10%Sn (coins, idols, bells, utencils etc.)

10. Constantan	60%Cu, 40%Ni (Electrical apparatus)
11. Common solder	50%Pb, 50%Sn
12. Coinage alloy	75%Cu, 25%Ni (For making coins)
13. Delta metal	Cu + Zn + Fe (Ship's propellers)
14. Duralumin	94.4%Al, 4%Cu, Mg. Mn, Si (For making air ships, pressure cookers)
15. Dutch metal	80%Cu, 20%Zn (Golden yellow, cheap ornaments)
16. Fine solder	56%Sn, 33%Pb
17. German sliver	60%Cu, 25%Zn, 15%Ni (Utencils)
18. Gun metal	86%Cu, 10%Sn, 4%Zn (For engineering works)
19. Invar	63%Fe, 36%Ni, 1%C (Watch pendulum)
20. Manganin	84%Cu, 12%Mn, 4%Ni
21. Mangnelium	85% to 99%Al, 1% to 15%Mg (Aeroplane's frame)

22. Monel metal	30%Cu, 70%Ni (For making alkali resistant containers)
23. Munz metal	60%Cu, 40%Zn (coins, tubes, castings)
24. Newtons' metal	Sn + Pb + Bi
25. Nicrome	Cr + Ni + Fe (Heater coil)
26. Pewter	75%Sn, 25%Pb (For metal soldering)
27. Phosphor bronze	85%Cu, 13%Sn, 2%P
28. Plumber's solder	70%Pb, 30%Sn
29. Rolled Gold	90%Cu, 10%Al (Cheap ornaments)
30. Rose's fusible metal	Bi + Pb + Sn (Automatic fuses, M.P. = 83°C)
31. Solder	Sn + Pb (For metal soldering)
32. Stainless steel	73%Fe, 1%C, 18%Cr, 8%Ni (For making automobile parts)
33. Stalloy	Fe + Si
34. Type metal	75%Pb, 20%Sb, 5%Sn (Compositor's Type)

35.	Wood's metal	Bi + Pb + Sn + Cd (For automatic fuses M.P. = 60°C)
36.	Y-alloy	Cu + Al
37.	Amatol	80%NH_4NO_3 + 20%T.N.T. (explosive)
38.	Gun powder	75%KNO_3, 12%S, 13% Charcoal
39.	Rectified spirit	95.6% ethyl alcohol, 4.4% Water
40.	Vinegar	10% acetic acid solution

17

IMPORTANT COMPOUNDS AND THEIR FORMULAE

1. Active nitrogen → Atomic nitrogen
2. Alums → $MAl(SO_4)_2.12H_2O$; (M = NH_4^+, Na^+, K^+ etc.)
3. Amatol → 80% NH_4NO_3 + 20% T.N.T. (explosive)
4. Anhydrone → $Mg(ClO_4)_2$
5. Aqua regia → conc. $1HNO_3$ + conc. 3HCl
6. Arsine → AsH_3
7. Asbestos → $CaMg_3(SiO_3)_4$
8. Borane → Hydrides of Boron
9. Bremstone → S_8
10. Blue vitriol → $CuSO_4.5H_2O$
11. Bleaching powder → Ca(OCl)Cl
12. Baryta water → $Ba(OH)_2$ Solution
13. Baryta → BaO

14. Baking powder or Soda → $NaHCO_3$
15. Black jack → Zinc ore
16. Calgon → $Na_2[Na_4(PO_3)_6]$
17. Carborundum → SiC
18. Caliche → Natural $NaNO_3$ containing $NaIO_3$
19. Caustic Soda → NaOH
20. Caustic potash → KOH
21. Calomel → Hg_2Cl_2
22. Cerussite → $PbCO_3$
23. Cementite → Fe_3C (iron carbide)
24. Chrom alum → $K_2SO_4.Cr_2(SO_4)_3.24H_2O$
25. Chinese white → ZnO
26. Corrosive sublimate → $HgCl_2$
27. D.D.T. → p-dichloro-diphenyl-trichloro-ethene
28. Deuterium → D or $_1H^2$ (Isotope of hydrogen)
29. Dry ice → Solid CO_2
30. Fehling's solution → A deep blue solution = $CuSO_4.5H_2O + NaOH +$ Na, K-tartrate (used for the test of aldehydes)
31. Feldspar → $KAlSi_3O_8$

32. Fenton's reagent → H_2O_2 + few drops of $FeCl_3$
33. Freon → CF_2Cl_2
34. Ferric alum → $K_2SO_4.Fe_2(SO_4)_3.24H_2O$
35. Fusion mixture → $Na_2CO_3 + K_2CO_3$
36. Fluid magnesia → 12% aqueous slon. of $Mg(HCO_3)_2$
37. Glauber's salt → $Na_2SO_4.10H_2O$
38. Graphite → An allotrope of carbon
39. Green vitriol → $FeSO_4.7H_2O$
40. Gun powder → 75% KNO_3, 12%S, 13% charcoal
41. Heavy hydrogen → D_2
42. Heavy water → D_2O
43. Hydrolith → CaH_2
44. Hypo → $Na_2S_2O_3.5H_2O$
45. Killed spirit → $ZnCl_2 + ZnO$ (Zn-oxy chloride)
46. Kesserite → $MgSO_4.H_2O$
47. Leuna saltpetre → Fertilizer [NH_4NO_3 + $(NH_4)_2SO_4$]
48. Lime or Quick lime → CaO
49. Lead of pencil → Graphite (C)

50. Lime water → A clear aqueous solution of $Ca(OH)_2$
51. Laughing gas → N_2O
52. Lunar caustic → $AgNO_3$
53. Litharge → PbO
54. Lithopone → A white pigment ($ZnS + BaSO_4$)
55. Massicot → PbO
56. Matte → $Cu_2S + FeS$
57. Magnesia alba → $2MgCO_3.Mg(OH)_2.3H_2O$
58. Magnesia → MgO
59. Marsh gas → Methane (CH_4)
60. Marble → $CaCO_3$
61. Micro cosmic salt → $NaNH_4HPO_4$ (used in the test of silicates)
62. Milk of magnesia → A paste of $Mg(OH)_2$ in water
63. Mohr's salt → $FeSO_4(NH_4)_2SO_4.6H_2O$
64. Muriatic acid → HCl
65. Milk of lime → Suspension of $Ca(OH)_2$ in water
66. Minium → Pb_3O_4
67. Nascent hydrogen → Atomic hydrogen
68. Nessler's reagent → Aq. soln. of K_2HgI_4

69. Nitro chalk → Fertilizer $[NH_4NO_3 + (NH_4)_2CO_3]$
70. Nitrolim → $CaCN_2$
71. Nitrophos → $Ca(H_2PO_4)_2 + 2Ca(NO_3)_2$
72. Oil of vitriol → conc. H_2SO_4
73. Ozone → O_3
74. Oleum → $H_2S_2O_7$
75. Oxygen gas → O_2
76. Pharaoh's Serpents → $Hg(CNS)_2$
77. Philosopher's Wool → ZNO
78. Phosphine → PH_3
79. Phosgene → $COCl_2$
80. Pig iron → impure form of iron
81. Potas alum → $K_2SO_4.Al_2(SO_4)_3.24H_2O$
82. Producer gas → A mixture of $(CO + N_2)$
83. Plaster of paris → $Ca(SO_4)_2.H_2O$
84. Quartz → $(SiO_2)_n$
85. Quick silver → Hg
86. Quick lime → CaO
87. Red lead → Pb_3O_4
88. Refrigerant → NH_3, CO_2, CF_2Cl_2 etc.
89. Rochelle Salt → Sodium potassium tartrate

90. Rust → $Fe_2O_3.xH_2O$
91. Sorel's Cement → $Mg(OH)Cl$
92. Soda-lime → $NaOH + CaO$
93. Soda ash or Sal soda → Na_2CO_3
94. Spathose ore → $FeCO_3$
95. Salammoniac → NH_4Cl
96. Slaked lime → $Ca(OH)_2$
97. Sal volatile (smelling salt) → $(NH_4)_2CO_3$
98. Spinel → $MgAl_2O_4$
99. Superphosphate → $Ca(H_2PO_4)_2 + 2CaSO_4$
100. T.N.T. → Tri-nitro toluene (explosive)
101. T.N.B. → Tri-nitro benzene (more powerful explosive than T.N.T.)
102. Tincal → $Na_2B_4O_7.10H_2O$
103. Talc → $3MgO.4SiO_2.H_2O$
104. Tritium → T or, H^3, an isotope of hydrogen
105. Vermilion → HgS (red)
106. Water glass → Sodium metasilicate (Na_2SiO_3)
107. Water gas → $CO + H_2$

108. Wrought iron → Pure form of iron
109. White vitriol → $ZnSO_4.7H_2O$
110. White lead → $2PbCO_3.Pb(OH)_2$
111. Zinc white → ZnO

18

ORGANIC COMPOUNDS

-1. **Organic Compounds:** There exist a large number of organic compounds. The property of carbon responsible to form so many compounds is due to following reasons:

1. **Tendency to form multiple bonds:** Carbon has covalency of four and forms multiple bonds (double bonds and triple bonds) with other carbon atoms, oxygen and nitrogen.
2. **Stability of bonds:** Carbon forms stable bonds with elements having electronegativity close to it.
3. **Catenation:** Ability of carbon to bond successively with other carbon atoms to form chains of varying lengths and shapes. This property of carbon is called as catenation and it is one of the main reasons of carbon to form so many compounds.

4. **Isomerism:** Compounds having same molecular formula but different structural formula are called isomers and this phenomenon is called as isomerism. Such as Molecular formula C_2H_6O represents two different compounds.

$CH_3 - CH_2 - OH$ (Ethyl alcohol) $\quad$ $CH_3 - O - CH_3$ (Dimethyl ether.)

2. **Hydrocarbons:** Hydrocarbons are the organic compounds having carbon atoms and hydrogen atoms only.

Classification of hydrocarbons

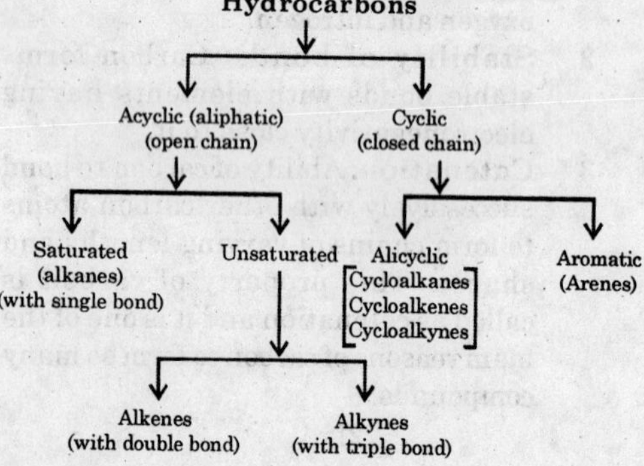

A. **Acyclic or aliphatic or open chain hydrocarbons :** These hydrocarbons contains open chain of carbon atoms in their molecules.

(a) **Saturated aliphatic hydrocarbons:** These compounds have carbon atoms linked to another by single bonds. Alkanes are saturated aliphatic hydrocarbons having general formula C_nH_{2n+2}. Such as methane (CH_4), ethane (C_2H_6), propane (C_3H_8) etc.

(b) **Unsaturated aliphatic hydrocarbons:** Unsaturated compounds have carbon atoms bonded either through double bond (in alkenes) or through triple bond (in alkynes).

Alkenes (C_nH_{2n})

```
  H   H
  |   |
H-C = C-H
```

Ethene (C_2H_4)

```
  H   H   H
  |   |   |
H-C = C - C-H
          |
          H
```

Propene (C_3H_6)

Alkynes (C_nH_{2n-2})

$$H-C\equiv C-H$$

Ethyne (C_2H_2)

$$H-\overset{1}{C}\equiv\overset{2}{C}-\overset{\overset{H}{|}\,3}{\underset{\underset{H}{|}}{C}}-H$$

Propyne (C_3H_4)

B. Cyclic or closed chain hydrocarbons : Cyclic hydrocarbons have their both the ends joined. These compounds may be saturated or unsaturated.

(a) Alicyclic: These are the cyclic hydrocarbons having properties similar to corresponding aliphatic hydrocarbons. These are classified into following types:

(*i*) ***Cycloalkanes:*** Cycloalkanes are saturated alicyclic hydrocarbons.

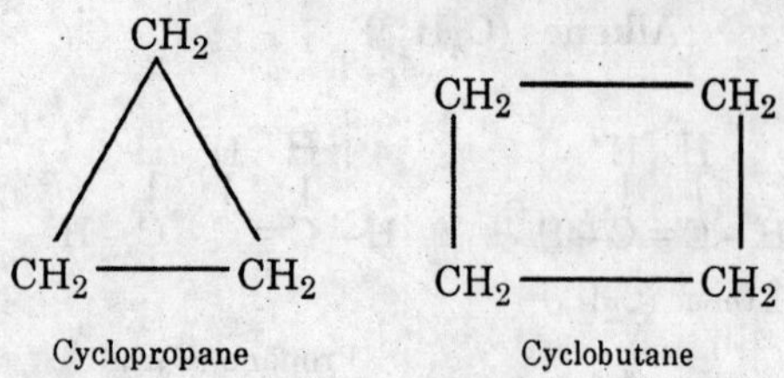

Cyclopropane Cyclobutane

CH_2

CH_2 CH_2

CH_2—CH_2

Cyclopentane

(*ii*) ***Cycloalkenes:*** Cycloalkenes are unsaturated hydrocarbons containing one carbon-carbon double bond.

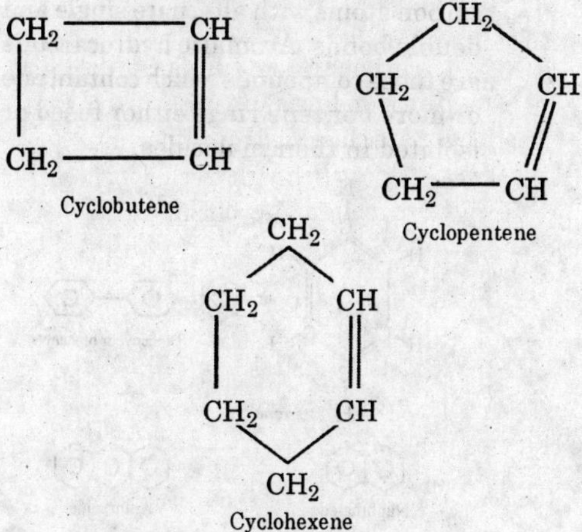

Cyclobutene

Cyclopentene

Cyclohexene

(iii) *Cycloalkynes:* Cycloalkynes are unsaturated hydrocarbons containing one carbon-carbon triple bond.

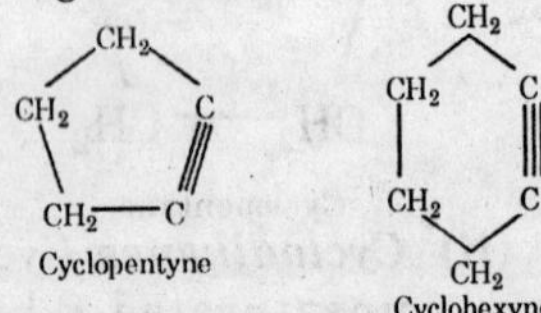

Cyclopentyne
Cyclohexyne

(b) Aromatic hydrocarbons: Benzene is the parent hydrocarbon having six carbon atoms with alternate single and double bonds. Aromatic hydrocarbons are those compounds which contain one or more benzene rings either fused or isolated in their molecules.

Benzene

Diphenyl or biphenyl

Naphthalene

Anthracine

3. **Homologous Series:** A homologous series is a class of organic compounds in which adjacent members differ by CH_2 unit. Individual member is called homologue and this phenomenon is called homology.
4. **Nomenclature of organic compounds:** Now-a-days, naming of organic compounds is done according to a set of rules laid down by International Union of Pure and Applied Chemistry (IUPAC), according to which, name of an organic compound has three parts :
   (i) Word root (ii) Suffix (iii) Prefix

(*i*) Word root: Word root signifies the number of carbon atoms in principal chain of organic compounds, as follows:

Chain length	Word root	Chain length	Word root
C_1	Meth–	C_{11}	Undec–
C_2	Eth–	C_{12}	Dodec–
C_3	Prop–	C_{13}	Tridec–
C_4	But–	C_{14}	Tetradec–
C_5	Pent–	C_{15}	Pentadec–
C_6	Hex–	C_{16}	Hexadec–
C_7	Hept--	C_{17}	Heptadec–
C_8	Oct–	C_{18}	Octadec–
C_9	Non–	C_{19}	Nonadec–
C_{10}	Dec–	C_{20}	Icosa–

(*ii*) Suffix

(a) ***Primary suffix:*** It indicated the saturated or insturated nature of principal carbon chain.

Principal carbon chain	Primary suffix
C – C (Saturated)	–ane
C = C (Unsaturated)	–ene
C ≡ C (Unsaturated)	–yne

If principal carbon chain has two, three or more double or triple bonds, following suffix are used.

Number of bonds	Double bond	Triple bond
2	– diene	– diyne
3	– triene	– triyne

(a) **Secondary suffix:** It indicated the type of functional group present in the organic compound.

Class of organic compounds	Functional group	Secondary suffix
Alcohols	– OH	– ol
Aldehydes	– CHO	– al
Ketones	> CO	– one
Carboxylic acids	– COOH	– oic acid
Esters	– COOR	alkyl ... oate

Acid amides	$-CONH_2$	– amide
Acid chlorides	$-COCl$	– oly chloride
Amines	$-NH_2$	– amine
Imines	$=NH$	– imine
Thio alcohols	$-SH$	– thiol
Nitriles	$-CN$	– nitrile

The terminal 'e' of primary suffix is dropped if secondary suffix starts with a vowel. However, if secondary suffix starts with a consonant, then the terminal 'e' of primary suffix is retained.

(iii) **Prefix** : Certain groups which are considered as substituents are indicated by prefixes. Prefixes are added immediately before the word root.

Substituent groups	**Prefix**	**Substituent groups**	**Prefix groups**
$-CH_3$	Methyl	$-F$	Fluoro
$-C_2H_5$	Ethyl	$-Cl$	Chloro
$-C_3H_7$	Propyl	$-Br$	Bromo
$-C_6H_5$	Phenyl	$-I$	Iodo
$-CH(CH_3)_2$	Iso-propyl	$-NO_2$	Nitro
$-C(CH_3)_3$	t-Butyl	$-NO$	Nitroso
$-OCH_3$	Methoxy	$-N=N-$	Diazo
$-OC_2H_5$	Ethoxy	$-OH$	Hydroxo

Hence complete IUPAC name of an organic compound can be written as follows:

Prefix + Word root + Primary suffix + Secondary suffix → IUPAC name

(i)

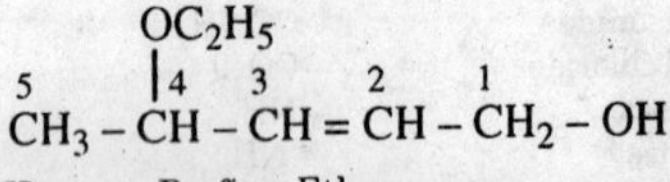

$$\overset{5}{CH_3} - \overset{4}{\underset{|}{CH}}(OC_2H_5) - \overset{3}{CH} = \overset{2}{CH} - \overset{1}{CH_2} - OH$$

Here, Prefix : Ethoxy
Word root : Pent –
Primary suffix : –ene
Secondary suffix : –ol

Hence, IUPAC name : 4–ethoxy–2–pentenol.

Nomenclature of Saturated Hydrocarbons (Alkanes)

Alkanes are those aliphatic hydrocarbons which contain single covalent bond between carbon atoms. Following rules are given below:

Rule1. Longest chain rule

Select the longest continuous chain, having maximum number of carbon atoms in the molecule. This chain is known as parent chain and word root is selected according to this chain. The other parts not included in the parent chain are considered as substituents.

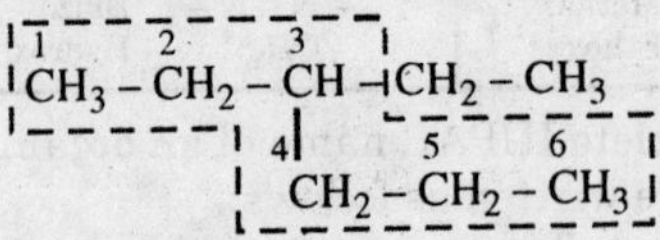

$$\overset{1}{CH_3} - \overset{2}{CH_2} - \overset{3}{CH} - CH_2 - CH_3$$
$$\quad\quad\quad\quad\quad | \; 4$$
$$\quad\quad\quad\quad\quad CH_2 - \overset{5}{CH_2} - \overset{6}{CH_3}$$

Correct chain
Chain Containing 6 C-atoms)

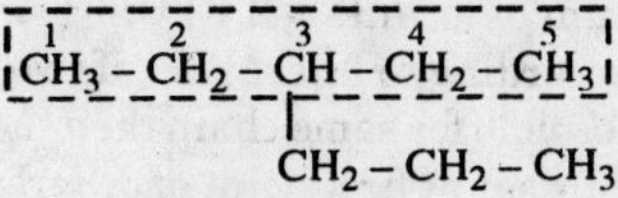

$$\begin{array}{ccccccccc} 1 & & 2 & & 3 & & 4 & & 5 \\ CH_3 & - & CH_2 & - & CH & - & CH_2 & - & CH_3 \\ & & & & | & & & & \\ & & & & CH_2 & - & CH_2 & - & CH_3 \end{array}$$

Incorrect chain
Chain Containing 5 C-atoms)

Rule 2. Lowest set of locants

Number the parent chain, starting from the end, in such a way that carbon atom linked to substituent gets the lowest possible number.

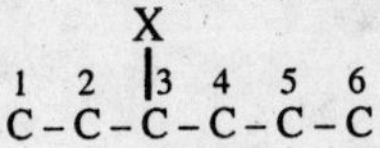

$$\begin{array}{ccccccccccc} & & & & X & & & & & & \\ 1 & & 2 & & |3 & & 4 & & 5 & & 6 \\ C & - & C & - & C & - & C & - & C & - & C \end{array} \qquad \begin{array}{ccccccccccc} & & & & X & & & & & & \\ 6 & & 5 & & |4 & & 3 & & 2 & & 1 \\ C & - & C & - & C & - & C & - & C & - & C \end{array}$$

Correct 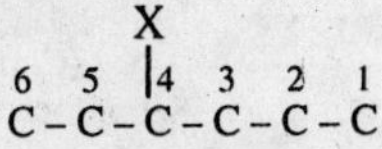 Incorrect

Rule 3. Lowest sum rule

If more than one substituents are attached to the longest continuous carbon chain, the numbering is done in such a way that sum of numbers given to substituents is the lowest.

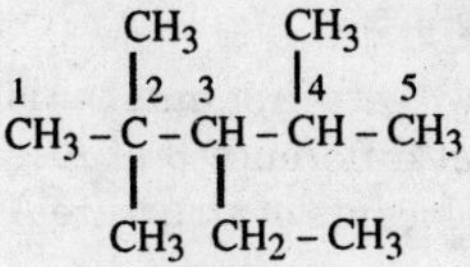

$$\begin{array}{ccccccccc} & & CH_3 & & & & CH_3 & & \\ 1 & & |2 & & 3 & & |4 & & 5 \\ CH_3 & - & C & - & CH & - & CH & - & CH_3 \\ & & | & & | & & & & \\ & & CH_3 & & CH_2 & - & CH_3 & & \end{array} \qquad \begin{array}{ccccccccc} & & CH_3 & & & & CH_3 & & \\ 5 & & 4| & & 3 & & |2 & & 1 \\ CH_3 & - & C & - & CH & - & CH & - & CH_3 \\ & & | & & | & & & & \\ & & CH_3 & & CH_2 & - & CH_3 & & \end{array}$$

(Correct) 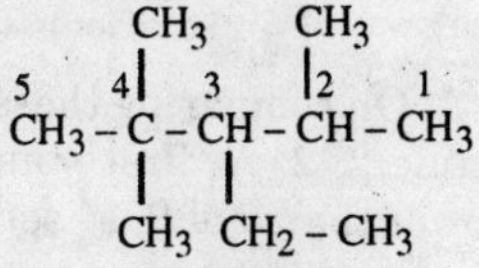 (Incorrect)

Sum of locants = 2 + 2 + 3 + 4 = 11

Sum of locants = 2 + 3 + 4 + 4 = 13

According to latest rule, if two sets of locants are possible for same chain then the set having lowest number at first point of difference is correct even if it violates the lowest sum rule. Such as,

$$\text{I. } \overset{1}{CH_3} - \underset{\ \ }{\overset{\overset{CH_3}{|}}{\overset{2}{CH}}} - \underset{\underset{C_2H_5}{|}}{\overset{3}{CH}} - \overset{4}{CH_2} - \underset{\underset{CH_3}{|}}{\overset{5}{CH}} - \overset{6}{CH_3}$$

(Correct)

Set of locants = 2, 3, 5

$$\text{II. } \overset{6}{CH_3} - \overset{\overset{CH_3}{|}}{\overset{5}{CH}} - \underset{\underset{C_2H_5}{|}}{\overset{4}{CH}} - \overset{3}{CH_2} - \underset{\underset{CH_3}{|}}{\overset{2}{CH}} - \overset{1}{CH_3}$$

(Incorrect)

Set of locants = 2, 4, 5

On comparing these two sets first number is same 2, 2. At first point of difference 3 of A is lower than 4 of B, so set of locants of structure A is correct.

$$\text{I.}\quad \overset{1}{C}H_3-\underset{\underset{CH_3}{|}}{\overset{\overset{CH_3}{|}}{\overset{2}{C}}}-\overset{3}{C}H_2-\underset{\underset{CH_3}{|}}{\overset{4}{C}H}-\overset{5}{C}H_3$$

(Correct)

Set of locants = 2, 2, 4

$$\text{II.}\quad \overset{5}{C}H_3-\underset{\underset{CH_3}{|}}{\overset{\overset{CH_3}{|}}{\overset{4}{C}}}-\overset{3}{C}H_2-\underset{\underset{CH_3}{|}}{\overset{2}{C}H}-\overset{1}{C}H_3$$

(Incorrect)

Set of locants = 2, 4, 4

Comparing these two sets first number is same 2, 2. At first point of difference (2, 4), 2 of structure A is lower than 4 of structure B, so set of locants of structure A is correct.

Rule 4. Naming of different substituents

(*a*) When two or more than two substituents are present on the parent chain, they are arranged in alphabetical order irrespective of their positional number.

$$\overset{1}{CH_3}-\underset{\displaystyle CH_3}{\overset{2}{CH}}-\underset{\displaystyle Br}{\overset{3}{CH}}-\overset{4}{CH_2}-\overset{5}{CH_2}$$

3- bromo-2-methylpentane (correct)

2-methyl-3-bromopentane (wrong)

(*b*) When two different substituents are present at equivalent positions, the substituent first in alphabetical order gets the lower number.

$$\overset{7}{CH_3}-\overset{6}{CH_2}-\underset{\displaystyle CH_3}{\overset{5}{CH}}-\overset{4}{CH_2}-\underset{\displaystyle CH_2CH_3}{\overset{3}{CH}}-\overset{2}{CH_2}-\overset{1}{CH_3}$$

3-ethyl-5-methyl heptane (correct)

5-ethyl-3-methyl heptane (wrong)

Rule 5.

For same substituents occuring more than once, perfixes di, tri, tetra, etc. are used. Which are not considered while deciding alphabetical order.

Rule 6.

If two possiblities are there for the same length of parent chain, number the compound so as to get maximum number of side chains or alkyl groups.

$$\begin{array}{l} \quad\;\; CH_3 \qquad\qquad\qquad\qquad\quad CH_2-CH_3 \\ ^{7}\quad\;\; |^{6} \quad\; ^{5} \qquad\; ^{4} \qquad\quad |^{3} \quad\; ^{2} \qquad ^{1} \\ CH_3-CH-CH_2-CH_2-CH-CH-CH_3 \\ \qquad\qquad\qquad\qquad\qquad\qquad\qquad\;\; | \\ \qquad\qquad\qquad\qquad\qquad\qquad\quad\; CH_3 \end{array}$$

(Correct)

3-ethyl-2, 6-dimethyl heptane (3 side chains)

$$\begin{array}{l} \quad\;\; CH_3 \qquad\qquad\qquad\qquad\quad CH_2CH_3 \\ \quad\;\; | \qquad\qquad\qquad\qquad\qquad\;\; | \\ CH_3-CH-CH_2-CH_2-CH-CH-CH_3 \\ \qquad\qquad\qquad\qquad\qquad\qquad\qquad\;\; | \\ \qquad\qquad\qquad\qquad\qquad\qquad\quad\; CH_3 \end{array}$$

(Incorrect)

5-isopropyl-2, 6-methyl heptane (2 side chains)

Rule 7.

When the substituent on the parent chain is complex, then it is numbered separately and carbon atom to which this substituent is linked to parent chain is numbered as 1. The name of complex substituent is written in bracket.

$$\overset{1}{CH_3} - \overset{2}{CH_2} - \overset{3}{CH_2} - \overset{4}{CH_2} - \overset{5}{CH} - \overset{6}{CH_2} - \overset{7}{CH_2} - \overset{8}{CH_2} - \overset{9}{CH_3}$$

$$\begin{array}{l} \quad | \\ {}^{1}CH - CH_3 \\ \quad | \\ {}^{2}CH - CH_3 \\ \quad | \\ {}^{3}CH_3 \end{array}$$

Complex substituent

5–(1, 2- dimethyl propyl) nonane

Nomenclature of Alkenes

The general rules for nomenclature of alkenes are exactly the same as for alkanes, with only following differences –

(*a*) Double bond should be in the longest continuous parent carbon chain.

(*b*) Numbering of carbon atoms should be done in such a way that double bond gets the lowest possible number.

(*c*) Suffix 'ane' is replaced by 'ene'.

Molecular formula	IUPAC name (Common name)
$CH_2 = CH_2$	Ethene (Ethylene)
$CH_3 - CH = CH_2$	1-Propene (Propylene)
$\overset{4}{CH_3} - \overset{3}{CH_2} - \overset{2}{CH} = \overset{1}{CH_2}$	1-Butene (α- Butylene)

$\overset{4}{CH_3} - \overset{3}{CH} = \overset{2}{CH} - \overset{1}{CH_2}$ 2-Butene (β-Butylene).

$\overset{1}{CH_2} - \overset{2}{\underset{}{C}}(CH_3) - \overset{3}{CH_2} - \overset{4}{CH_3}$ 2-methyl-1-butene.

$\overset{1}{CH_2} - \overset{2}{C}(CH_3) = \overset{3}{CH} - \overset{4}{CH}(CH_3) - \overset{5}{CH_3}$ 2,4-dimethyl-2-pentene

$CH_3 - CH_2 - \overset{2}{C}(={}^{1}CH_2) - \overset{3}{CH_2} - \overset{4}{CH_2} - \overset{5}{CH_2} - \overset{6}{CH_3}$ 2-ethyl-2-hexene.

$CH_3 - CH_2 - CH_2 - CH_2 - \overset{4}{C}(\overset{3}{CH_2} - \overset{2}{CH} = \overset{1}{CH_2}) = \overset{5}{CH} - \overset{6}{CH_3}$ 4-butyl-hexa-1, 4-diene

$\overset{4}{CH_3} - \overset{3}{C}(CH_3) = \overset{2}{CH} - \overset{1}{CH_2}(Br)$ 1-bromo-3-methyl-2-butene

$\overset{1}{CH_3} - \overset{2}{CH} = \overset{3}{C} = \overset{4}{CH} - \overset{5}{CH_2} - \overset{6}{CH_3}$ 2, 3-hexadiene.

$\overset{5}{CH_2} = \overset{4}{CH} - \overset{3}{CH}(CH_2 - CH_3) - \overset{2}{C}(Cl) = \overset{1}{CH_2}$ 2-chloro-3ethyl-1, 4-pentadiene

Nomenclature of Alkynes

The general rules for nomenclature of alkynes are exactly the same as for alkanes, with only following differences:

(*a*) Triple bond should be in the longest continuous parent carbon chain.

(*b*) Numbering of carbon atoms should be done is such a way that triple bond gets the lowest possible number.

(*c*) Suffix 'ane' is replaced by 'yne'.

Molecular formula	IUPAC name (Common name)
$CH \equiv CH$	Ethyne (Acetylene)
$CH_3 - C \equiv CH$	Propyne (Methyl acetylene)
$\overset{4}{C}H_3 - \overset{3}{C}H_2 - \overset{2}{C} \equiv \overset{1}{C}H$	1-Butyne (Ethyl acetylene)
$\overset{4}{C}H_3 - \overset{3}{C} \equiv \overset{2}{C} - \overset{1}{C}H_3$	2-Butyne (Dimethyl acetylene)
$CH \equiv C - CH_2 - CH_2 - Cl$	4-chloro-1-butyne.
$\overset{1}{H}C \equiv \overset{2}{C} - \overset{3}{C}H(CH_3) - \overset{4}{C} \equiv \overset{5}{C}H$	3-methyl penta-1, 4-diyne
$\overset{1}{H}C \equiv \overset{2}{C} - \overset{3}{C}H = \overset{4}{C}H - \overset{5}{C} \equiv \overset{6}{C}H$	Hex-3-ene-1, 5-diyne

$$\overset{6}{CH_3}-\underset{\underset{CH_3}{|}}{\overset{5}{CH}}-\overset{4}{C}\equiv\overset{3}{C}-\underset{\underset{CH_3}{|}}{\overset{\overset{CH_3}{|}}{\overset{2}{C}}}-\overset{1}{CH_3}$$

2,2,5-trimethyl-3-hexyne

5. **Isomerism:** Compounds which have the same molecular formula but different chemical or physical properties are called as isomers and this phenomenon is called as isomerism.

Types of isomerism

(*a*) Structural isomerism.

(*b*) Stereoisomerism.

(*a*) Structural isomerism: If isomerism is due to difference in the arrangement of atoms within the molecule, the phenomenon is called structural isomerism.

Types of structural isomerism

(*i*) Chain isomerism: Chain isomers have the same molecular formula but differ in the order in which the carbon atoms are bonded to each other.

(*ii*) Position isomerism: Position of a functional group on the carbon atom differs in position isomerism.

(*iii*) **Functional isomerism:** Functional isomers have the same molecular formula but different functional groups.

(*iv*) **Metamerism:** Unequal distribution of carbon atoms on either side of the functional group results in this type of isomerism.

(*v*) **Tautomerism:** In this isomerism, isomers are in dynamic equilibrium with each other.

(*b*) **Stereoisomerism:** Isomerism caused due to its different arrangements of atoms in space is called as stereoisomerism.

Types of stereoisomerism

(*i*) ***Geometrical isomerism:*** Geometrical isomerism results from a restriction in rotation about double bonds.

(*ii*) ***Optical isomerism:*** Atoms of optical isomers have different spatial arrangement. Optical isomers have the ability to rotate plane polarised light.

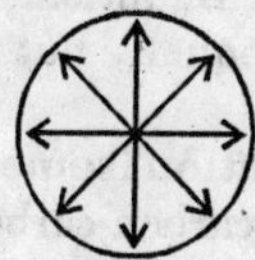

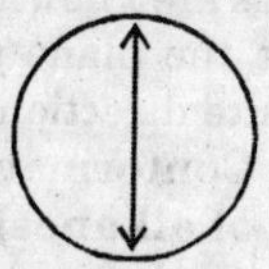

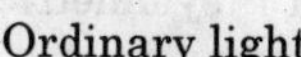

Ordinary light

Plane polarised light

6. **Optical activity and chairality:** Those compounds which have ability to rotate the plane of polarised light are optically active. This property of a compound is called **optical activity**.

The compound which rotates the plane of polarised light to the right, is called dextro-rotatory. The compound which rotates the plane of polarised light to the left, is called laevo-rotatory.

Plane of symmetry: A plane which divides a compound into two symmetrical halves, is known to be the plane of symmetry.

Chirality: The chiral atoms do not have any plane of symmetry and are not superimposable on its mirror image. This property is known as chirality.

Enantiomers: Non-superimposable mirror images are known as enantiomers. Enantiomers rotate the plane polarised light equally but in opposite directions.

7. **Conformations (Conformation isomers of alkanes):** Alkanes have carbon-carbon single sigma (σ) bond that is symmetrical about internuclear axis and is capable of free rotation. The various arrangements of atoms in a molecule that may arise due to free rotation about a single bond are known as conformations of the molecule. The conformational forms are as follows:
   (*i*) **The eclipsed form:** In eclipsed form atoms are as close to each other as possible. The hydrogen atoms bonded with back carbon atoms are eclipsed by hydrogen atoms bonded with front carbon atoms.
   (*ii*) **The staggered form:** In staggered form atoms are as far as possible.
   (*iii*) **The skew form:** Except from these two forms there are a large number of possible arrangements in between these two forms which are known as skew forms.

Preparation of Alkanes

I. By hydrogenation of unsaturated hydrocarbons

$$\underset{\text{Alkene}}{R-CH=CH_2} + H_2 \xrightarrow[523\ -\ 573\ K]{\text{Ni or Pt or Pd}} \underset{\text{Alkane}}{R-CH_2-CH_3}$$

$$\underset{\text{Ethene}}{CH_2=CH_2} + H_2 \xrightarrow[523\ -\ 573\ K]{\text{Ni or Pt or Pd}} \underset{\text{Ethane}}{CH_3-CH_3}$$

$$\underset{\text{Alkyne}}{R-C\equiv CH} + 2H_2 \xrightarrow[523\ -\ 573\ K]{\text{Ni or Pt or Pd}} \underset{\text{Alkane}}{R-CH_2-CH_2}$$

$$\underset{\text{Ethyne}}{CH\equiv CH} + 2H_2 \xrightarrow[523\ -\ 573\ K]{\text{Ni or Pt or Pd}} \underset{\text{Ethane}}{CH_3-CH_3}$$

II. From alkyl halides

(*a*) By Wurtz reaction:

$$\underset{\text{Alkyl halide}}{R-\boxed{X+2Na+X}-R} \xrightarrow{\text{Dry ether}} \underset{\text{Alkane}}{R-R} + 2NaX$$

(Sodium)

$$\underset{\text{Bromomethane}}{CH_3-Br} + 2Na + Br-CH_3 \xrightarrow{\text{Dry ether}} \underset{\text{Ethane}}{CH_3-CH_3} + 2NaBr$$

Methane cannot be prepared by Wurtz reaction.

(*b*) By reduction with nascent hydrogen:

$$\underset{\text{Alkyl halide}}{R-X} + \underset{\text{Nascent hydrogen}}{2[H]} \rightarrow \underset{\text{Alkane}}{R-H} + HX$$

$$CH_3CH_2Br + 2[H] \rightarrow CH_3{-}CH_3 + HBr$$

Bromoethane — Ethane

III. From aldehydes and ketones

$$CH_3COCH_3 + 2H_2 \xrightarrow{Zn+conc.HCl} CH_3CH_2CH_3 + H_2O$$

Propanone — Propane

$$CH_3CHO + 2H_2 \xrightarrow{Zn+conc.HCl} CH_3CH_3 + H_2O$$

Ethanal — Ethane

IV. From carboxylic acid

(a) By decarboxylation (Removal of CO_2)

$$R-COONa + NaOH \xrightarrow{CaO,\ 630\ K} R-H + Na_2CO_3$$

Sodium carboxylate — Sodium hydroxide — Alkane

$$CH_3-COONa + NaOH \xrightarrow{CaO,\ 630\ K} CH_4 + Na_2CO_3$$

Sodium acetate — Sodium hydroxide — Methane

Calcium oxide keeps sodium hydroxide dry. In laboratory, methane is prepared by this method.

(b) By electrolysis of sodium or potassium salts of carboxylic acid.

$$2RCOONa + 2H_2O \xrightarrow{Electrolysis} R-R + 2NaOH + H_2 + 2CO_2$$

Sodium carboxylate — Alkane (at anode)

$$2CH_3COONa + 2H_2O \xrightarrow{Electrolysis} CH_3{-}CH_3 + 2NaOH + H_2 + 2CO_2$$

Sodium ethanoate — Ethane

Properties of Alkanes

I. **Boiling points:** Boiling points of alkanes increase with the increase in their molecular mass. Branched alkanes have lower boiling points than corresponding straight chain alkanes.

II. **Melting points:** Alkanes do not show regular variation in melting points. Alkanes with even number of carbon atoms have higher melting points than with odd number of carbon atoms.

III. **Density:** Densities of alkanes increase with increase in molecular mass. Alkanes are lighter than water.

IV. **Solubility:** Hydrocarbons are soluble in non-polar solvents (benzene, carbon tetrachloride, petroleum, ether etc.) and insoluble in polar solvents (water, alcohols etc.)

V. **Combustion**

$$CH_4 + 2O_2 \rightarrow CO_2 + 2H_2O + \text{Heat}$$

While in limited supply of air CO and H_2O are obtained.

$$2CH_4 + 3O_2 \rightarrow 2CO + 4H_2O$$

VII. Substitution reactions:

(a) Halogenation :

$$CH_4 + Cl_2 \xrightarrow{\text{Sunlight}} CH_3Cl + HCl$$

$$CH_3Cl + Cl_2 \rightarrow \underset{\text{Dichloro methane}}{CH_2Cl_2} + HCl$$

$$CH_2Cl_2 + Cl_2 \rightarrow \underset{\text{Trichloro Methane (Chloroform)}}{CHCl_3} + HCl$$

$$CHCl_3 + Cl_2 \rightarrow \underset{\text{Tetrachloro methane}}{CCl_4} + HCl$$

The order of reactivity of halogens with alkanes is $F_2 > Cl_2 > Br_2 > I_2$.

(b) Sulphonation :

$$\underset{\text{Alkane}}{R-H} + \underset{\text{Sulphuric acid (fuming)}}{HO-SO_3H} \xrightarrow{\Delta} \underset{\text{Alkane sulphonic acid}}{R-SO_3H} + H_2O$$

Where $R = C_6H_{13}$ or larger alkyl group.

(c) Nitration :

$$CH_4 + HNO_3 \xrightarrow{400-500°C} CH_3NO_2 + H_2O$$

(d) Isomerization:

$$\underset{\text{n-butane}}{CH_3CH_2CH_2CH_3} \xrightarrow{AlCl_3 + HCl,\ 250°\ C} \underset{\text{Isobutane}}{CH_3-\overset{\overset{CH_3}{|}}{CH}-CH_3}$$

VI. Aromatization:

$$C_6H_{14} \xrightarrow[600^\circ C,\ 15\ atm]{Cr_2O_3\ /\ Al_2O_3} C_6H_6\ (\text{Benzene}) + 4H_2$$

Preparation of Alkenes

I. By dehydration of alcohols:

$$\underset{\text{Primary alcohol}}{R-CH_2-CH_2OH} \xrightarrow[\text{or } Al_2O_3,\ 623-633\ K]{\text{Conc. } H_2SO_4,\ 440K} \underset{\text{Alkene}}{R-CH=CH_2} + H_2O$$

$$\underset{\text{Ethanol}}{CH_3CH_2OH} \xrightarrow[\text{or } Al_2O_3,\ 623-633\ K]{\text{Conc. } H_2SO_4,\ 440K} \underset{\text{Ethene}}{CH_2=CH_2} + H_2O$$

II. By cracking of alkanes:

$$\underset{\text{Ethane}}{CH_3-CH_3} \xrightarrow{600^\circ C} \underset{\text{Ethene}}{CH_2=CH_2} + H_2$$

III. By controlled hydrogenation of alkynes:

$$\underset{\text{Alkyne}}{R-C\equiv C-H} + H_2 \xrightarrow[\text{quinoline}]{\text{Pd-CaCO}_3} \underset{\text{Alkene}}{R-CH=CH_2}$$

$$\underset{\text{Propyne}}{CH_3-C\equiv C-H} + H_2 \xrightarrow[\text{quinoline}]{\text{Pd-CaCO}_3} \underset{\text{Propene}}{CH_3-CH=CH_2}$$

IV. By dehydrohalogenation of alkyl halides:

$$\underset{\text{Alkyl halide}}{R\text{–}CH_2\text{–}CH_2X} + \underset{\text{Potassium hydroxide}}{KOH} \xrightarrow[353-363\,K]{(\text{Alcoholic})} \underset{\text{Alkene}}{R\text{–}CH=CH_2} + KX + H_2O$$

$$\underset{\text{Iodoethane}}{CH_3CH_2\text{–}I} + \underset{\text{Potassium Hydroxide (alc.)}}{KOH} \xrightarrow{353-363\,K} \underset{\text{Ethene}}{CH_2=CH_2} + KI + H_2O$$

$$\underset{\text{1-bromopropane}}{CH_3CH_2CH_2-Br} + \underset{\text{Potassium hydroxide (atc.)}}{KOH} \xrightarrow{353-363\,K} \underset{\text{Propene}}{CH_3CH=CH_2} + KBr + H_2O$$

V. By dehalogenation of vic-dihalides (having two halogen atoms on adjacent carbon atoms.)

$$\underset{\text{Dibromo-alkane}}{R-\underset{Br}{\underset{|}{CH}}-\underset{Br}{\underset{|}{CH_2}}} \xrightarrow[\Delta]{Zn/C_2H_5OH} \underset{\text{Alkene}}{R-CH=CH_2} + Br_2$$

$$\underset{\text{1, 2-dibromoethane}}{\underset{Br}{\underset{|}{CH_2}}-\underset{Br}{\underset{|}{CH_2}}} \xrightarrow[\Delta]{Zn/C_2H_5OH} \underset{\text{Ethene}}{CH_2=CH_2} + Br_2$$

$$\underset{\text{1, 2-dibromopropane}}{CH_3-\underset{Br}{\underset{|}{CH}}-\underset{Br}{\underset{|}{CH_2}}} \xrightarrow[\Delta]{Zn/C_2H_5OH} \underset{\text{Propene}}{CH_3-CH=CH_2} + Br_2$$

Properties of Alkenes

I. **State:** Alkenes having carbon atoms upto 4 are gases, having 5-15 carbon atoms are liquids, but members having more than 15 carbon atoms are solids.

II. **Boiling points and melting points:** Alkenes have higher boiling and melting points than corresponding alkanes. The boiling points and melting points of alkenes also increase with increase in molecular mass like alkanes.

III. **Combustion:**

$$CH_2=CH_2 + 3O_2 \rightarrow 2CO_2 + 2H_2O + \text{Heat}$$

IV. **Oxidation:** With cold and dilute potassium permanganate solution ($KMnO_4$).

$$\underset{\text{Ethene}}{CH_2=CH_2} + H_2O + [O] \xrightarrow{\text{Neutral } KMnO_4} \underset{\text{Ethylene glycol}}{\underset{OH\quad\;\; OH}{CH_2-CH_2}}$$

$$CH_3CH=CHCH_3 + H_2O + [O] \xrightarrow{KMnO_4} \underset{\text{2, 3-butanediol}}{CH_3-\overset{OH}{CH}-\overset{OH}{CH}-CH_3}$$

V. Addition reactions: Addition of hydrogen:

$$\underset{\text{Alkene}}{R-CH=CH_2} + H_2 \xrightarrow[475-525\ K]{\text{Ni or Pt}} \underset{\text{Alkane}}{R-CH_2-CH_3}$$

$$\underset{\text{Ethene}}{CH_2=CH_2} + H_2 \xrightarrow[475-525\ K]{\text{Ni or Pt}} \underset{\text{Ethane}}{CH_3-CH_3}$$

VI. Polymerization:

$$\underset{\text{Ethene}}{n\,CH_2=CH_2} \xrightarrow{473-673\ K,\ \text{High pres.}} \underset{\text{Polyethene}}{(-CH_2-CH_2-)_n}$$

$$\underset{\text{Chloro ethane (Vinyl chloride)}}{n\,CH_2=CH-Cl} \longrightarrow \underset{\text{Polyvinyl chloride (PVC)}}{(-CH_2-\underset{Cl}{\underset{|}{C}H_2}-)_n}$$

$$\underset{\text{Styrene}}{n\,\underset{C_6H_5}{\underset{|}{H}C}=CH_2} \longrightarrow \underset{\text{Poly styrene}}{(-\underset{C_6H_5}{\underset{|}{C}H}-CH_2-)_n}$$

$$\underset{\text{Tetrafluoroethane}}{n\,CF_2=CF_2} \longrightarrow \underset{\text{Polytetrafluoroethane (Teflon)}}{(-CF_2-CF_2-)_n}$$

VII. Isomerization:

$$\underset{\text{1-butene}}{CH_3-CH_2+CH=CH_2} \xrightarrow{AlCl_3,\ 500\text{-}700°C} \underset{\text{2-butene}}{CH_3-CH=CH-CH_3}$$

Preparation of Alkynes

I. From ethyne (for higher alkynes):

$$\underset{\text{Ethyne}}{HC \equiv CH} + \underset{\text{Sodamide}}{NaNH_2} \xrightarrow{196\text{ K, Liq. }NH_3} \underset{\text{Sodium acetylide}}{HC \equiv C^-Na^+} + NH_3$$

$$HC \equiv C^-Na^+ + \underset{\text{Bromomethane}}{CH_3-Br} \rightarrow \underset{\text{Propyne}}{HC \equiv C-CH_3} + NaBr$$

$$HC \equiv C^-Na^+ + \underset{\text{Bromoethane}}{CH_3-CH_2-Br} \rightarrow \underset{\text{1-Butyne}}{HC \equiv C-CH_2CH_3} + NaBr$$

II. From calcium carbide:

$$CaC_2 + 2H_2O \rightarrow \underset{\text{Ethyne}}{HC = CH} + Ca(OH)_2$$

Properties of Alkynes

I. M.P., B.P, solubility and density of alkynes are similar to those of alkanes.

II. **Oxidation:** With cold potassium permanganate solution:

$$\underset{\text{Ethyne}}{3CH \equiv CH} + \underset{\text{Potassium permanganate (Pink)}}{4KMnO_4} + 2H_2O \xrightarrow{298-303\text{ K}} \underset{\text{1, 2-Ethanedione}}{3H-\overset{\overset{\displaystyle O}{\|}}{C}-\overset{\overset{\displaystyle O}{\|}}{C}-H} + \underset{\text{Manganese dioxide}}{4MnO_2} + 4KOH$$

III. Addition reaction: Addition of hydrogen.

$$\underset{\text{Alkyne}}{R-C\equiv CH} \xrightarrow{H_2/Ni} \underset{\text{Alkene}}{R-CH=CH_2} \xrightarrow{H_2/Ni} \underset{\text{Alkane}}{RCH_2CH_3}$$

$$\underset{\text{Ethyne}}{HC\equiv CH} \xrightarrow{H_2/Ni} \underset{\text{Ethene}}{CH_2=CH_2} \xrightarrow{H_2/Ni} \underset{\text{Ethane}}{CH_3-CH_3}$$

$$\underset{\text{Propyne}}{CH_3C\equiv CH} + H_2 \xrightarrow[\text{Lindlar's catalyst}]{Pd/CaCO_3 + \text{lead acetate}} \underset{\text{Propene}}{CH_3-CH=CH_2}$$

IV. Acidic character of alkynes: Hydrogen bonded with carbon atom having triple bond is acidic in nature.

(a) Formation of heavy metal acetylides:

$$\underset{\text{Acetylene}}{H-C\equiv C-H} + \underset{\text{Silver nitrate}}{2AgNO_3} + \underset{\text{Ammonium hydroxide}}{2NH_4OH} \rightarrow \underset{\substack{\text{Silver acetylide}\\\text{(white ppt)}}}{Ag-C\equiv C-Ag} + 2NH_4NO_3 + 2H_2O$$

$$\underset{\text{Acetylene}}{H-C\equiv C-H} + \underset{\text{Cuprous chloride}}{Cu_2Cl_2} + 2NH_4OH \rightarrow \underset{\substack{\text{Cuprous acetylide}\\\text{(red ppt)}}}{Cu-C\equiv C-Cu} + 2NH_4Cl + 2H_2O$$

(b) Formation of alkali metal cynides :

$$\underset{\text{Acetylene}}{H-C\equiv C-H} + Na \rightarrow \underset{\text{Sodium acetylide}}{H-C\equiv C-Na} + H$$

$$\underset{\text{Acetylene}}{H-C\equiv C-H} + \underset{\text{Sodamide}}{NaNH_2} \rightarrow \underset{\text{Sodium acetylide}}{HC\equiv C-Na} + NH_3$$

Some Important Facts

I. Alkanes are also called as **paraffins.**

II. Alkanes and cycloalkanes show conformational isomerism.

III. Pi (π) bond lacks symmetry about internuclear axis.

IV. Pi (π) bond is weaker than sigma (σ) bond.

V. Cis-form has substituent on the same side and trans-form has substituent on opposite side.

VI. Alkenes show position and the chain isomerism, as well as geometrical isomerism.

VII. Alkynes show position and chain isomerism but do not show geometrical isomerism.

VIII. For geometrical isomerism, the molecule must have a double bond.

IX. Ethyne does not show isomerism.

X. Bond length decreases in following order: $C-C > C=C > C\equiv C$

XI. **Reforming:** It is the process of conversion of alicyclic or aliphatic hydrocarbons having 6-8 carbon atoms into aromatic hydrocarbons.

XII. **Cracking:** It is the process of conversion of higher hydrocarbons into lower hydrocarbons.

XIII. **Cetane number:** Cetane number of a diesel fuel is defined as the percentage of cetane in a mixture of centane and α-methyl naphthalene which has the same ignition qualities as the test fuel.

XIV. **Octane number:** Octane number of a fuel is defined as the percentage of iso-octane in the mixture of n-heptane and iso-octane which has the same anti-knocking property as the test fuel.

XV. **Carbonization** is the process of transformation of organic matter into coal in absence of oxygen.

XVI. Gaseous mixture found with crude petroleum is known as **natural gas**.

XVII. Petroleum is also known as **rock oil**.

XVIII. Branched chain compounds, alkenes, cycloalkanes and aromatic compounds have high octane numbers.

XIX. Methane is the chief constituent of natural gas.

XX. Thermal cracking is difficult to control.

XXI. Lead compounds are poisonous and responsible for air pollution.

XXII. Acidic nature of alkynes can be used to separate, to purify and to identify them from other hydrocarbons.

XXIII. Alkymides are generally explosive and unstable in dry state.

XXIV. Wurtz reaction is not suitable for preparation of unsymmetrical alkanes.

XXV. On addition of H_2O, only C_2H_2 gives aldehyde and rest of all alkynes give ketones.

XXVI. Alkymides are ionic in nature.

19

CARBOHYDRATES, ENZYMES AND NUCLEIC ACID

The cell: Cell is the smallest fuctional and structural unit of all living matters. The cell is composed of C, H, N, O, P and S etc.

Structure of Cell

(*a*) **Cell wall:** It is the regid wall around the cell made up of carbohydrates such as cellulose, lignin etc.

(*b*) **Cell membrane:** It is the elastic membrane next to cell wall made up of phospholipids and proteins.

(*c*) **Cytoplasm:** It is the viscous translucent jelly like material having organelles embedded in it.

(*d*) **Mitochondria:** It is releases ATP by food oxidation during respiration.

(*e*) **Lysosomes:** It contains enzymes to digest food, foreign substances and worn out organelles.

(*f*) **Nucleus:** It is the control house of the cell having DNA packed with histone protein in it.

(*g*) **Endoplasmic reticulum:** It has ribosomes on its outer surface made up of r-RNA employed in protein synthesis.

(*h*) **Golgi apparatus:** It delivers biomolecules such as phospholipids, proteins etc. to other organelles.

Components of a bacterial cell

	Components		Percent of total cell weight
I.	Water		70
II.	Protein		15
III.	Nucleic acids:	RNA	6
		DNA	1
IV.	Polysaccharides		2
V.	Phospholipids		2
VI.	Inorganic ions (Na^+, K^+, Mg^{2+} Ca^{2+})		1
VII.	Miscellaneous small molecules		3

Carbohydrates

Carbohydrates are the polyhydroxyaldehydes or polyhydroxyketones, and all those substances which yield these compounds on hydrolysis. They have the general formula $C_n(H_2O)_n$.

Classification of Carbohydrates

(*a*) **Monosaccharides:** These are the single unit of carbohydrates having general formula $C_nH_{2n}O_n$ (n = 3 to 8). These are sweet, water soluble, optically active and non-hydrolysable sugars. Sugars containing aldehyde group are known as **aldoses** and those containing keto group are known as **ketoses**. Some examples are as follows:

	Type of sugar	Formula	Aldose	Ketose
I.	Trioses	$C_3H_6O_3$	Aldotriose (Glycer-aldehyde)	Ketotriose (Di-hydrox-yacetone)
II.	Tetroses	$C_4H_8O_4$	Aldotetrose (Erythrose)	Ketotetrose (Eruthrulose)
III.	Pentoses	$C_5H_{10}O_5$	Aldopentose (Ribose)	Ketopentose (Xylulose)
IV.	Hexoses	$C_6H_{12}O_6$	Aldohexose (Glucose)	Ketohexose (Fructose)

(b) Oligosaccharides: Oligosaccharides are the sugars composed of 2-10 monosaccharides. Those containing two simple sugars are known as **disaccharides** and those having three simple sugar are known as **trisaccharides**. Sugars belonging to this class are sweet, water soluble and hydrolysable.

(i) Sucrose ($C_{12}H_{22}O_{11}$, table sugar): It is dextro-rotatory, disaccharide of glucose and fructose. The two monosaccharides are joined together by α-1, 2-glycoside bond. Under mild acidic conditon, on hydrolysis, it yields equimolar mixture of glucose and fructose which is called as invert sugar.

$$\underset{\text{Sucrose}}{C_{12}H_{22}O_{11}} + H_2O \xrightarrow[\text{or Invertase}]{H^+} \underset{\text{Glucose}}{C_6H_{12}O_6} + \underset{\text{Fructose}}{C_6H_{12}O_6}$$

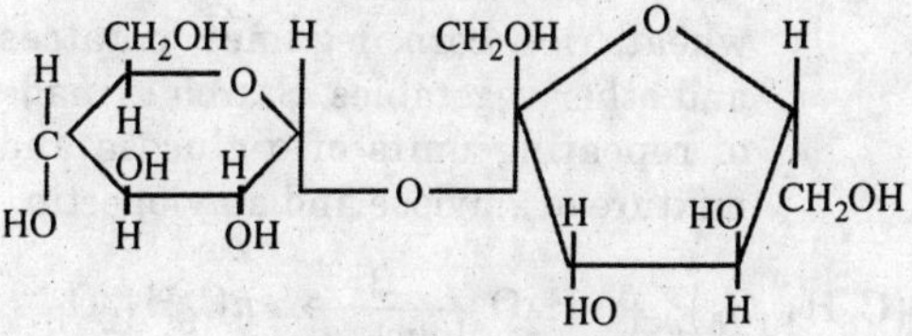

Structure of sucrose

***(ii)* Maltose ($C_{12}H_{22}O_{11}$, malt sugar):** It is a disaccharide obtained by partial hydrolysis of starch by the enzyme diastase. On hydrolysis under acidic conditions it yields two molecules of D(+) glucose.

$$\underset{\text{Maltose}}{C_{12}H_{22}O_{11}} + H_2O \xrightarrow{H^+} \underset{\text{Glucose}}{C_6H_{12}O_6} + \underset{\text{Glucose}}{C_6H_{12}O_6}$$

***(iii)* Lactose ($C_{12}H_{22}O_{11}$, milk sugar):** It is a disaccharide of glucose and galactose.

$$\underset{\text{Lactose}}{C_{12}H_{22}O_{11}} + H_2O \xrightarrow{H^+} \underset{\text{Glucose}}{C_6H_{12}O_6} + \underset{\text{Galactose}}{C_6H_{12}O_6}$$

(c) Plysaccharides: Polysaccharides contain more than ten units of monosaccharides.

***(i)* Starch $[C_6H_{10}O_5)_n]$** : It occures in roots, tubers and seeds. Large amount (60 – 80%) of starch is present in wheat, rice, corn, legumes, potatoes and other vegetables. Starch is made of repeating units of α-glucose and mixture of amylose and amylopectin.

$$(C_6H_{10}O_5)_n + nH_2O \xrightarrow[\text{Hydrolysis}]{H^+} \underset{\text{Glucose}}{nC_6H_{12}O_6}$$

(*ii*) **Cellulose:** It is most abundant of all carbohydrates and is the chief constituent of wood (50%), cotton and paper etc. It is a linear polymer of β-glucose.

(*iii*) **Glycogen:** It is the animal polysaccharide and is stored in liver and muscles. It contains 10-14 units of glucose.

Functions of Carbohydrates

(*a*) They provide energy for the functioning of living system.

(*b*) They serve as the structural materials for cell walls.

(*c*) Glycogen carbohydrate stored in liver acts as instant food on hydrolysis glucose.

2. **Proteins:** It may be defined as condensation polymers of α-amino acids.

Amino acids: These are the bifunctional compounds containing amino acid and a carboxylic acid group.

$$\underbrace{H_2N}_{\text{Amino group}} - \overset{\displaystyle R}{\underset{\displaystyle H}{\overset{|}{\underset{|}{C}}}} - \underbrace{COOH}_{\text{Carboxylic acid group}}$$

Peptide bond: Amino acids are joined through peptide bonds. peptide bonds are formed by condensation of carboxylic acid of an amino acid with amino group of the same or another molecule of amino acid.

$$\underset{\text{Carboxylic group}}{-\overset{\displaystyle O}{\overset{\|}{C}}-OH} + \underset{\text{Aminogroup}}{H-\underset{\displaystyle H}{\underset{|}{N}}-} \rightarrow \underset{\text{Peptide bond}}{-\underset{\displaystyle O}{\underset{\|}{C}}-\underset{\displaystyle H}{\underset{|}{N}}-} + H_2O$$

Functions of Proteins

(*i*) They are essential for growth and maintenance of life.

(*ii*) They are structural material of animal tissues like skin, hair, nails etc.

(*iii*) They act as transport agents like haemoglobin.

3. **Enzymes**

   (*i*) e.g. Amylase, invertase, lactase, maltase, urease etc.

Enzymes are the powerful biological catalyst. They are all proteins, higly efficient (at pH = 7 and temperature = 37°C) and reaction specific. Some enzymes are associated with non-protein part known as **co-enzymes**.

Enzymes and the reactions they catalyse

	Enzyme	Reaction catalysed
I.	Amylase	Starch → Glucose
II.	Carbonic anhydrase	Carbonic acid → $H_2O + CO_2$.
III.	Invertase	Sucrose → Glucose + Fructose
IV.	Lactase	Lactose → Glucose + Galactose
V.	Maltase	Maltose → Glucose
VI.	Pepsin	Protein → Amino acids
VII.	Urease	Urea → $NH_3 + CO_2$

The dificiency of an enzyme can cause serious disease. The dificiency of an enzyme phenylalanine causes the disease phenyl ketone urea; dificiency of enzyme tyrosinase cause albinism.

4. Nucleic acids

e.g. Deoxyribonucleic acid (DNA) and ribonucleic acid (RNA).

Nucleic acids are the polynucleotide which are composed of sugar, phosphoric acid group and four bases, two purines and two pyremidines. The two type of nucleic acid are deoxyribonucleic acid (DNA) and ribonucleic acid (RNA).

• **Nucleoside**

	Sugar	**Base**		**Acid**
		Purines	**Pyrimidines**	
RNA	Ribose	Adenine (A)	Cytosine (C)	Phosphoric acid
		Guanine (G)	Uracil (U)	
DNA	Dexyribose	Adenine (A)	Cytosine (C)	Phosphoric acid
		Guanine (G)	Thymine (T)	

Nucleic acids transmit hereditary effects from one generation to next generation and also cotntrol biosynthesis of protein.

Double Helix

Nucleic acid DNA has two polynucleotide chains held together by hydrogen bonds and twisted about a common axis to form double helix. The two strands of double helix are complementary. A thymine base (T) is linked to adenine base (A) in the opposite chain and guanine base (G) is always linked to cytosine base (C) in the opposite chain.

5. **Viruses:** Chemically, viruses contain nucleic acid surrounded by simple proteins. They are ultramicroscopic infectious agents and are responsible for many diseases

known as **virus diseases**. e.g. AIDS (caused by HIV), common cold, rabies, yellow fever, poliomyelitis, measles etc.

6. **Lipids:** These are oily substances, insoluble in water and soluble in benzene, chloroform etc.

   (i) **Triglycerides:** These are the triesters of unbranched long chain saturated or unsaturated fatty acids with glycerol. Triglycerides containing higher proportion of saturated fatty acids are known as fats, while those with higher proportion of unsaturated fatty acids are known as **oils**.

Some Important Facts

I. Aldehydric and ketonic groups in carbohydrates exist in combination with one hydroxyl group of the molecule in the form of hemiacetals and hemiketals.

II. Glucose is also known as grape sugar, blood sugar and corn sugar.

III. In maltose, two D (+) glucose molecules are joined together by α–1, 4–glycoside bond.

IV. In lactose, D(+) glucose and D(+) galactose are joined together by β–1, 4–glycoside bond.

V. Humans cannot digest cellulose because of absence of enzymes to hydrolyse it.

VI. Glucose exists in two optically active forms, α–D–Glucose and β–D–glucose.

VII. Pentoses and hexoses are most common naturally occurring monosaccharides.

VIII. Sucrose is dextrorotatory but on hydrolysis it yields equimolar solution of glucose and fructose (invert sugar) that is levorotatory. This inversion of optical rotation is called as **inversion of sugar**.

IX. More the unsaturated acid, lower is the melting point of fat.

X. Fats differ from each other due to different acid radicals present in triglycerides.

XI. Proteins are nitrogenous organic compounds of high molecular masses.

XII. Protein is name given to a polypeptide containing more than 100 amino acid molecules.

XIII. DNA contains the genetic code and directs protein synthesis through RNA.

XIV. Mutation is any physical or chemical change that alters the sequence of bases in DNA molecule.

XV. Amylopectin is a branched chain polymer and has molecular mass in the range of 50,000 – 100,000 amu.

XVI. Insulin is a harmone which contains 51 amino acids in two polypeptide chains.

XVII. The enzymes are much larger than the molecules they catalyse.

XVIII. Anything that causes mutation is called a **mutagen.**

20

POLYMERS

Polymers. They are high molecular mass compounds made up of a large number of simple repeating units known as monomers.

Polymers synthesized from one type of monomers are known as **homopolymers** such as polythene, which has one type of monomer ethylene.

Polymers synthesized from two or more types of monomers are known as **co-polymers** such as terylene, which has two types of monomers, ethylene glycol and terephthalic acid.

Classification of Polymers Based on Source

(*a*) **Natural polymers:** The polymers that are found in nature are known as natural polymers. Strach is a polymers of glucose, protein is a polymer of α-amino acids and natural rubber is a polymer of 2-methyl-1, 3-butadiene (isoprene).

$$nCH_2 = \underset{\underset{\text{Isoprene}}{CH_3}}{\underset{|}{C}} - CH = CH_2 \xrightarrow{\text{Polymerization}} \left[- CH_2 - \underset{CH_3}{\underset{|}{C}} = CH - CH_2 - \right]$$

Polyisoprene (natural robber)

(*b*) **Synthetic polymers:** The polymers prepared in the laboratories are known as synthetic polymers. Such as polyethylene, polystyrene, bakelite, nylon etc.

Some Important Facts

I. All polymers are macromolecules but all macromolecules are not polymers.

II. The addition polymers have the same empirical formula as their monomers.

III. Carbohydrates and proteins are biopolymers.

IV. Metaphosphoric acid $(HPO_3)_n$, silicates and silicones are some inorganic polymers.

V. Natural rubber is obtained from white milky juice called latex of rubber trees.

VI. Thiokol rubber is made by polymerization of ethylene dichloride and sodium polysulphide.

Some important polymers and their monomers

Polymers	Monomers	Structural formula
Addition polymers		
I. Polyethylene or Polythene	Ethene	$(-CH_2-CH_2-)_n$
II. Polystyrene	Styrene	$[-CH(C_6H_5)-CH_2-]_n$
III. Polypropylene or polypropene	Propylene	$(-CH(CH_3)-CH_2-)_n$
IV. Buna-S	1, 3- butadiene and styrene	$(-CH_2-CH=CH-CH_2-CH(C_6H_5)-CH_2-)_n$
V. Neoprene	Chloroprene	$(-CH_2-C(Cl)=CH-CH_2-)_n$
VI. Polyacrylonitrile (PAN) or Orlon	Vinyl cyanide	$(-CH_2-CH(CN)-)_n$

VII.	Polyvinyl chloride (PVC)	Vinyl chloride	$(-CH_2-\underset{\underset{Cl}{\mid}}{CH}-)_n$
VIII.	Polytetrafluoroethylene (PTFE) or Teflon	Tetrafluoroethylene	$(F_2C-CF_2-)_n$
Condensation polymers			
IX.	Nylon-6	Caprolactum	$(-NH-(CH_2)_5-\overset{\overset{O}{\parallel}}{C}-)_n$
X.	Nylon-66	Hexamethylenediamine and adipic acid	$(-\overset{\overset{N}{\mid}}{N}-(CH_2)_6-\overset{\overset{H}{\mid}}{N}-\overset{\overset{O}{\parallel}}{C}-(CH_2)_4-\overset{\overset{O}{\parallel}}{C}-)_n$
XI.	Glyptal	Ethylene glycol and phthalic acid	$(-O-CH_2-CH_2-O-\overset{\overset{O}{\parallel}}{C}-C_6H_4-\overset{\overset{O}{\parallel}}{C}-)_n$
XII.	Terylene or Dacron	Ethylene glycol and and terephthalic acid	$(-OCH_2-CH_2-\overset{\overset{O}{\parallel}}{C}O-C_6H_4-\overset{\overset{O}{\parallel}}{C}-)_n$

VII. Rayon is a man-made fibre obtained from cellulose, cotton etc.

VIII. Silk is obtained from cocoons of silk worms.

IX. Silk is a natural fibre which consists of fibre protein (fibroin) and gummy protein (sericin).

X. Cotton is obtained from cotton plants and it is mainly cellulose which is a polymer of β-glucose.

XI. Wool is an animal fibre which contains keratin (a protein) and grease (mainly cholesterol).

CHEMISTRY IN ACTION

Dyes

These are coloured chemical substances used to impart colours to the textile fibres, food stuffs, plastic materials, silk, wool, paper, leather and other materials.

Classification of Dyes

A. Classification of dyes based on application

I. **Acid dyes** : These dyes are usually sodium salts of sulphonic acid ($-SO_3H$) or phenolic compounds. These are used to colour wool, natural silk and nylon. They cannot be used to colour cotton. *Examples* : martius yellow, naphthol blue, methyl red, methyl orange, orange I, orange II etc.

$$Na^+O_3^-S-C_6H_4-N^+\equiv NCl^- + C_{10}H_7-OH \rightarrow$$

Diabolized sulphanilic acid

a-naphthol

$$Na^+O_3^-S-C_6H_4-N=N-C_{10}H_6-OH$$

Orange-I

II. **Direct dyes** : It can be applied by immersing fibre in hot aqueous solution of dye. These are suitable for fabrics which can form hydrogen bonds like nylon, rayon, cotton, wool and silk. *Example* : martius yellow, congo red, direct black etc.

III. **Basic dyes** : Basic dyes contain amino group and are used to dye nylons and polyesters. *Examples* : malachite green, aniline yellow, methyl violet etc.

IV. **Ingrain dyes** : These dyes are produced within the fabric by coupling of phenol or naphthol adsorbed on the surface of fabric with adiazonium salt. These are used to dye silk, cotton, polyester and nylon. *Examples* : azo dyes.

V. **Mordant dyes :** These dyes cannot be used directly and require another substance called as mordant to bind fabric and dye. A metal ion is used as mordant, *Example* : alizarin dye.

VI. **Fibre reactive dye :** These dyes form permanent chemical bonds with hydroxy or amino group of the fibres of cotton, wool or silk. Examples : derivatives of 2, 4-dichloro-1, 3, 5-triazine.

VII. **Vat dyes :** These dyes are applied to the fabric in reduced state in which dye is soluble and colourless and then oxidized to insoluble coloured dye. *Example* : indigo.

VIII. **Disperse dyes :** These are water insoluble dyes and are dispersed by suitable reagent before application on synthetic fibre. These are usually meant for nylon, polyester and polyacrylonitrile. Examples : anthraquinone dye, mono-azo dye etc.

B. Classification of dyes based on chemical constitution

These dyes are as follows :

Name of dye	Structural unit	Example
I. Azo dyes	$-N=N-$ Azo group	$Na^+O_3^-S-C_6H_4-N=N-C_{10}H_6-OH$ Orange I
II. Indigoid dyes	Indigoid group	Indigo
III. Phthalein dyes	Phthalein group	Phthalein group
IV. Nitro dyes	$-NO_2$ Nitro group	Martius yellow

V. **Anthraquinone dyes**	Anthraquinone group	Alizarin
VI. **Triphenyl methane dyes**	Triphenyl methane group	$N(CH_3)_2$ ... $= N^+ (CH_3)_2$ Malachite green

Medicines or drugs are chemical substances which are used to prevent or cure diseases.

I. **Antipyretics** : These are the substances used to bring down body temperature in high fever such as, aspirin, paracetamol and phenacetin.

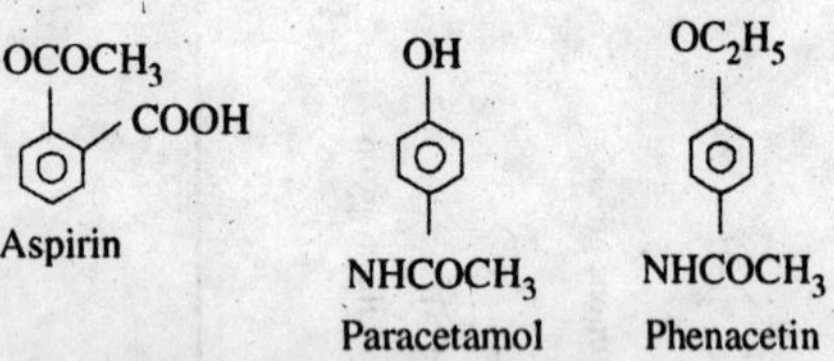

II. **Antiseptics** : These are applied on living tissues to kill or prevent the growth of micro-organisms, such as, tincture of iodine, dettol, bithional used in medicated soaps, genation violet, emthylene blue etc.

III. **Analgesics** : These are the substances used to relives pains, such as, aspirin, novalgin etc. Certain narcotics like morphine, marijuane, codeine and heroin are also used as analgesics.

IV. **Antibiotics** : These are the substances produced by micro-organisms to destroy or inhibit the growth of other micro-organisms, such as, penicillin is used for bronchitis, pneumonia, sore throat and access etc. Streptomycin is used for tuberculosis and tetramycin is used in typhoid fever, meningitis and local infections etc.

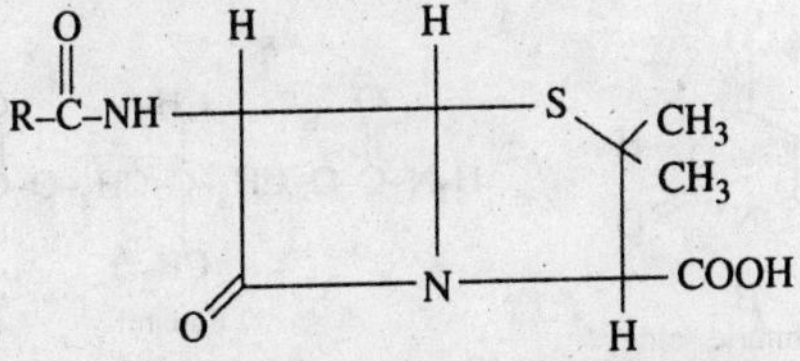

General structure of penicillin

V. Disinfectants : These are also used to kill micro-organisms while it cannot be applied on living tissues e.g. phenol.

VI. Sulpha drugs : Sulphadiazine, sulphanilamide, sulphathiozole, sulphaguanidine, sulphaacetamide etc. are sulpha drugs. They possess powerful antibactrial property.

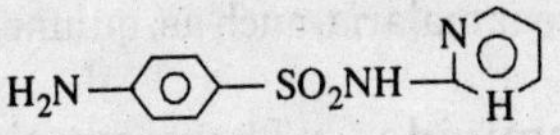

Sulphadiazine

VII. Tranquilizers : These are used to cure mental diseases like mental tension, depression etc. They act on higher centres of nervours system and are the constituents of sleeping pills, such as, barbituric acid and its derivativs like luminal and seconal.

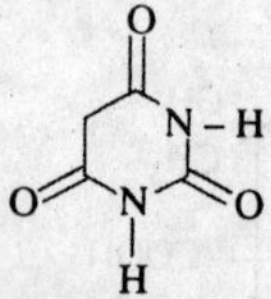

Barbituric acid

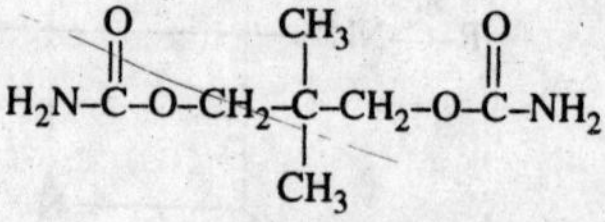

$$H_2N-\overset{\overset{O}{\|}}{C}-O-CH_2-\underset{\underset{CH_3}{|}}{\overset{\overset{CH_3}{|}}{C}}-CH_2-O-\overset{\overset{O}{\|}}{C}-NH_2$$

Equanil

Equanil is used in depression and hypertension.

VIII. Anaesthetics : These are used to produce general or local insensibility to pains and other sensations. Cocaine and novocaine are commonly used local anaesthetics. Chloroform, diethyl and divinyl ethers, nitrous oxide etc. are general anaesthetics.

IX. Antimalarial drugs : These drugs are used to treat malaria, such as, qunine, chloroquine etc.

X. Germicides : These are the chemical substances used to kill germs, virus and fungi, such as, phenol, cresols, formaldehyde, DDT, $KMnO_4$ solution (1%), bleaching powder, chlorine, hydrogen peroxide etc.

Rocket Propellants

A propellant is a combination of an oxidizer and fuel that when ignites undergoes combustion to release large quantities of hot gases. The passage of hot gases through the nozzle of the rocket motor gives necessary thrust for the rocket to move forward.

Types of Rocket Propellants

I. **Solid propellants :** The two types of solid propellants are :

(*a*) **Composite propellants :** These are composed of fuel such as polyurethane or polybutadiene, oxidizer suchas ammonium perchlorate, nitrate or chlorate and some additive like finely divided aluminium or magnesium metal.

(*b*) **Double base propellants :** They consist of nitroglycerine and nitrocellulose. Their ignition cannot be regulated.

II. **Liquid propellants** : The two types of liquid propellants are :

(*a*) **Monopropellants** : These contain single chemical compound that acts as a fuel as well as oxidiser, such as hydrazine, hydrogen peroxide, nitromethane, methyl nitrate etc.

(*b*) **Bipropellants** : These contain liquid fuel and liquid oxidizer separately and are allowed to combine if required such as kerosene and liquid oxygen, N_2O_4 and unsymmetrical dimethyl hydrazine (UDMH) and N_2O_4 and monomethyl hydrazine (MMH), liquid H_2 and liquid O_2 etc.

III. **Hybrid propellants** : It consist of solid fuel and liquid oxidizer, such as acrylic rubber (solid fuel) and liquid N_2O_4 (liquid oxidizer).

Some Important Facts

I. Alizarin (red) and indigo (blue) were probably the earliest dyes. These are obtained from plants.

II. Chromophore are responsible for selective absorption of light.

III. The dye consists of chromogen and auxochrome.

IV. Azo dyes are the largest class of synthetic dyes.

V. Vat dyes are water insoluble dyes so applied to the fabric in reduced state.

VI. Auxochrome normally deepens the colour of a chromogen.

VII. Imortant manufacturing units of penicilline are Hindustan Antibiotics Ltd, Pimpri and Indian Drugs and Pharmaceuticals Ltd. Rishikesh.

VIII. The first antibiotic was pencillin, it was discovered by Alexander Fleming.

CHEMO CURIOS

Here are described some important informative clues about the main chemical elements of the Periodic Table and chemistry as a whole. These are regarded as rememberable records of chemistry.

1. Most electropositive element: *Cs* (among stable elements)
2. Most electropositive element which is radioactive in nature: *Fr*
3. Most electronegative element: Fluorine ($En = 4.00$)
4. The second most electronegative element: Oxygen ($En = 3.5$)
5. Most conductive metal : Silver (Ag)
6. Most conductive non-metal : Graphite (element allotrope of carbon)
7. The only liquid metal at room temperature: Mercury (Hg, Z = 80)
8. Chemically most reactive non-metal: Fluorine

9. The only liquid non-metal : Br
10. Most poisonous element : Plutonium (Pu)
11. Element having lowest I. P. : Cs (among stable elements) I.P. : 3.89 *ev/atom*
12. Element having highest I.P.: He (I.P. = 24.58 *ev/atom*)
13. Element having highest electron affinity : Cl (Ea = 3.61 *ev/atom*)
14. Element having highest electron affinity next to chlorine : Fluorine (Ea = 3.45 *ev/atom*)
15. Least electropositive element: **Fluorine**
16. Platinum (Pt) is called : **White gold**
17. Mercury (Hg) is called : **Quick silver**
18. Petroleum is called : **Liquid gold**
19. 24 carat gold is called : **Pure gold**
20. Graphite is called : **Plumbago or black lead**
21. Graphite is used as **dry lubricant.**
22. Element kept in water : White phosphorus (P_4)
23. Elements kept in kerosene oil : Na, K, Rb, Cs
24. Element sublimes on heating : Iodine
25. Substances sublime on heating (*i*) Nepthalene (*ii*) Iodine (*iii*) Ammonium chloride (*iv*) Pthalic acid (*v*) Camphor etc.

26. Hydrogen is the only element in the P.T. whose nucleus contains no neutron.
27. The most inflamable gas is hydrogen.
28. Hydrogen is the sole element whose isotopes have separate names and symbols ($_1H^1$: Protium, $_1H^2$: Deuterium, $_1H^3$: Tritium)
29. Approximately 90% of the sun's mass is H_2.
30. Analysis of light emitted by stars indicates that most stars are predominantly hydrogen.
31. In interstitial hydrides hydrogen is present in atomic state. They occupy the vacant spaces of metallic structure.
32. There are only two elements (H and He) in the P.T. that contains zero core electron.
33. Radon (*Rn*, $Z = 86$) is the sole inert gas which is radioactive in nature.
34. Non-metals having metallic lusture : Iodine and Graphite.
35. Non-metal which is conductor of electricity : Graphite.
36. Hardest natural occuring substance : **Diamond.**
37. **Trans uranic** or man made elements : After $_{92}U$ i.e. 93 .... onwards.

38. Heaviest natural occuring element: ${}_{92}U$
39. All metals containing block : *d* and *f*-blocks.
40. Metals, non-metals and metalloids containing block : *p*-block
41. Typical metalloid elements : B, Si, Ge, As and Te
42. Lightest metalloid : **Boron** (3.30 g/cc)
43. Heaviest metalloid : **Tellurium** (6.23 g/cc)
44. Amphoteric metals : Zn, Al, Sn, Pb etc.
45. Elements showing diagonal relationship :

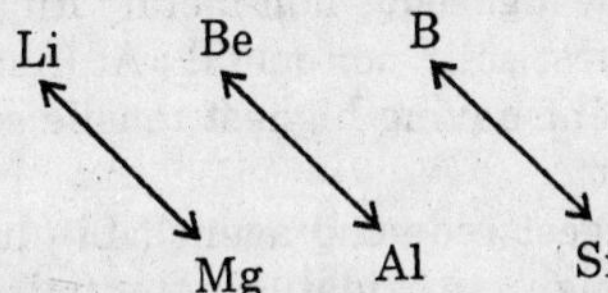

46. **Noble metals :** Au, Pt, Hg, Ag etc.
47. Element having smallest atomic size : Hydrogen
48. Largest atomic size : Cs (among stable elements)
49. Largest cation : Cs^+ (among stable elements)s
50. Largest anion : At^-
51. Smallest anion : F^-
52. A gaseous element having minimum b.p. : Helium

53. Non-metal having highest m.p. and b.p.: Diamond (due to giant covalent structure)
54. Metal having highest m.p. and b.p. : Tungusten (*W*)
55. Carbon has sp^3 hybridisation in diamond
56. Carbon has sp^2 hybridisation in graphite
57. Lightest known element: Hydrogen
58. Lightest solid metal : Li
59. Heaviest solid metal : Os (Osmium)
60. Lightest gaseous non-metal : Hydrogen
61. Heaviest gaseous non-metal : Rn (Radon)
62. Heaviest solid non-metal : At (Astatine)
63. Element having highest tensile strength: Boron
64. Strongest reducing agent : Li (due to its very high +ve oxidation potential)
65. Weakest oxidising agent : I_2 (among stable halogens)
66. Strongest halogen halide reducing agent : HI
67. Most electrovalent compound : CsF
68. Most covalent compound : Diatomic molecules (as H_2, Cl_2 etc.)
69. Most stable carbonate : Cs_2CO_3
70. Strongest base : CsOH
71. Strongest basic oxide : Cs_2O

72. All metaloxides are basic except : ZnO, PbO, Al_2O_3. (amphoteric oxides)
73. All non-metal oxides are acidic except : CO, NO, N_2O (neutral) and H_2O (amphoteric)
74. P_2O_3 and P_2O_5 are solid non-metallic acidic oxides.
75. Natural explosive: NCl_3
76. Artificial explosive : Dynamite
77. Solid CO_2 is called : **Dry Ice**
78. Dry bleacher : Ozone (O_3)
79. Oldest known halogen element : Chlorine
80. Latest known halogen element : Astatine (*At*)
81. Oldest known inert gas : Ar
82. Latest known inert gas : Radon (Rn)
83. *Ortho* and *para* hydrogens are isomers of hydrogen.
84. The most abundant element in the earth's crust : **Oxygen** (49.2% by weight)
85. The second most abundant element in the earth's crust : **Silicon** (25.7% by weight)
86. The third most abundant element in the earth's crust : **Aluminium** (8.1% by weight)
87. The most abundant gas in the atmosphere : **Nitrogen** (78% by volume approx.)
88. Rarest element of the earth's crust : **Astatine** (At)

89. Red variety of HgS is called : **Vermilion**
90. HgS is soluble in aqua-regia.
91. CaF_2 is insoluble in water
92. AgBr is soluble in conc. NH_4OH solution
93. AgI is insoluble in NH_4OH solution
94. $HgCl_2$ (white) is soluble in aqua-regia
95. AgCl (white) is soluble in dilute NH_4OH due to complex formation $[Ag(NH_3)_2]Cl$
96. Hydrolysis by means of an acid is called : **Acidolysis**
97. The process of separation of gases based on the difference in the rate of diffusion is called : **Atmolysis**
98. Thermal decomposition of organic compounds is called : **Pyrolysis**
99. Pyrolysis of alkane is called : **Cracking**
100. Azimuthal quantum number is otherwise called secondary or subsidiary or angular momentum quantum number.
101. Nuclear fission is the basis for the manufacture of **atom bomb**.
102. Nuclear fusion is the basis for the manufacture of **hydrogen bomb**
103. Helium (He) is found in the solar atmosphere

104. C^{14} – isotope is used in **carbon dating process** to determine the age of old wood, rock etc.
105. The man made element made in the first nuclear reactor was : Plutonium
106. Natural radioactivity is always an exothermic process
107. Radioactivity is a first order reaction.
108. The time for complete decay of a given sample of radio-element is practically infinity.
109. All elements after the atomic number 83 are radioactive.
N.B.: Elements with atomic number 43 and 61 are also radioactive.
110. D_2O (heavy water) is used as coolant in nuclear reactors.
111. Polonium has **27 isotopes**, more than any other element.
112. **Positron** ($_{+1}e^{0}$) is the **anti particle** of electron ($_{-1}e^{0}$)
113. Gamma ray has got no charge and no mass.
114. β-particle is equivalent to electron
115. α-particle ($_{2}He^{4}$) is equivalent to helium nucleus.
116. Atomic weights of almost all the elements are fractional. (Try to explain)

117. 0.529 Å is called the one atomic unit of length and is equal to **Bohr radius**.
118. Inside an atom, the energy of electron is always negative. (Try to explain)
119. At infinity, the energy of electron is zero.
120. Particle nature of electron is supported by photoelectric effect experiment.
121. Electron has dual (particle as well as wave) nature.
122. **Stern-Gerlach** experiment provides an experimental proof of the fact that angular momentum of electron is quantized.
123. Valency of an element is always +ve and whole number.
124. The **Vander Waal's radius** (non bonded radius) of an element is always greater than the **covalent radius**.
125. Co-ordinate bond is otherwise called **dative bond** or **co-ionic bond** or **semi-polar bond**.
126. Ionic compounds do not exhibit space isomerism.
127. Fe (iron) in solid state has both electrostatic and covalent bonds.
128. Molecule formed by like atoms but polar: O_3 (ozone)

129. Compound containing polar bond but is non-polar: CO_2 (due to its linear structure)
130. It seems that the oxidation number of sulphur in $H_2S_2O_8$ (peroxy di sulphuric acid) is +7, but actually it is +6 (due to presence of two peroxide linkages)
131. Interstitial hydrides are non-stoichiometric, because its composition changes with temperature and pressure
132. The oxidation number of N in N_2 is zero but, its valency is 3. (Try to explain)
133. Valency of C in $C_{12}H_{22}O_{11}$ is +4 but, its oxidation number is zero.
134. **Alchemy:** is chemistry of the middle ages, the chief aim of which was to discover how to change ordinary metal into gold.
135. **Amalgam:** is an alloy with mercury as one of the metals.
136. **Salinometer:** is an instrument for measuring the salinity of a solution.
137. Platinum is called: **Adam's Catalyst**
138. **Albamine** is the old name for astatine.
139. Lead acetate $[Pb(CH_3COO)_2]$ is called : **Sugar of lead** or **INORGANIC SALT**
140. H_2SO_4 is known as **Oil of vitriol** or **Battery acid**

141. An explosive mixture of T.N.T and NH_4NO_3 is called : **AMATOL**
142. An explosive mixture of NH_4NO_3 and Al-metal powder is called : **AMMONAL**
143. **Anthracite** is a variety of coal of high quality.
144. The lowest rank of coal is called : **Lignite**
145. Conc. HNO_3 is also called : **Aqua Fortis**
146. Peroxy disulphuric acid ($H_2S_2O_8$) is called : **Marshal's acid**
147. Fuming sulphuric acid ($H_2S_2O_7$) is also called : **Oleum** or **Nordhausen acid**
148. Hydrocyanic acid (HCN) is also called : **Prussic acid**
149. Peroxy mono sulphuric acid (H_2SO_5) is also called : **Caro's acid**
150. K_2CO_3 is called : **Potash** or **PEARL ASH**
151. $CaCO_3$ is called : **Iceland spar**
152. Deposits of impure $CaCO_3$ is called : **Coral**
153. $Ca_3(PO_4)_2$ is called : **Bone ash**
154. Animal charcoal is called : **Bone black**
155. **BATH SALT** : Na_2CO_3. $NaHCO_3.2H_2O$
156. **HAIR SALT** : $Al_2(SO_4)_3.18H_2O$ (alunogenite)
157. **Brunswick green**: $CuCl_2.3Cu(OH)_2$
158. Green vitriol ($FeSO_4.7H_2O$) is also called: **Copperas**
